PROTECTING THE PEOPLE

APPLYING GIS

PROTECTING THE PEOPLE

GIS FOR LAW ENFORCEMENT

Edited by
John Beck
Matt Artz

Esri Press
REDLANDS | CALIFORNIA

Esri Press, 380 New York Street, Redlands, California 92373-8100
Copyright © 2022 Esri
All rights reserved.

ISBN: 9781589487291
Library of Congress Control Number: 2022936933

The information contained in this document is the exclusive property of Esri or its licensors. This work is protected under United States copyright law and other international copyright treaties and conventions. No part of this work may be reproduced or transmitted in any form or by any means, electronic or mechanical, including photocopying and recording, or by any information storage or retrieval system, except as expressly permitted in writing by Esri. All requests should be sent to Attention: Director, Contracts and Legal Department, Esri, 380 New York Street, Redlands, California 92373-8100, USA.

The information contained in this document is subject to change without notice.

US Government Restricted/Limited Rights: Any software, documentation, and/or data delivered hereunder is subject to the terms of the License Agreement. The commercial license rights in the License Agreement strictly govern Licensee's use, reproduction, or disclosure of the software, data, and documentation. In no event shall the US Government acquire greater than RESTRICTED/LIMITED RIGHTS. At a minimum, use, duplication, or disclosure by the US Government is subject to restrictions as set forth in FAR §52.227-14 Alternates I, II, and III (DEC 2007); FAR §52.227-19(b) (DEC 2007) and/or FAR §12.211/12.212 (Commercial Technical Data/Computer Software); and DFARS §252.227-7015 (DEC 2011) (Technical Data–Commercial Items) and/or DFARS §227.7202 (Commercial Computer Software and Commercial Computer Software Documentation), as applicable. Contractor/Manufacturer is Esri, 380 New York Street, Redlands, California 92373-8100, USA.

Esri products or services referenced in this publication are trademarks, service marks, or registered marks of Esri in the United States, the European Community, or certain other jurisdictions. To learn more about Esri marks, go to: links.esri.com/EsriProductNamingGuide. Other companies and products or services mentioned herein may be trademarks, service marks, or registered marks of their respective mark owners.

For purchasing and distribution options (both domestic and international), please visit esripress.esri.com.

On the cover: Photograph by Keith Homan.

CONTENTS

Introduction — vii
How to use this book — xi

PART 1: ANALYSIS — 1

Smart policing gets a boost from enterprise GIS — 3
St. Petersburg Police Department

Collaborating to address crime that arrests alone won't solve — 9
Philadelphia Police Department

Sharing maps, apps, and dashboards to improve public safety — 16
St. Louis Metropolitan Police Department

Enhancing strategic planning and crime analysis with data analytics — 20
Altoona Police Department

PART 2: OPERATIONS — 27

Saving time and money by increasing field data collection efficiency — 29
Redlands Police Department

Purpose-driven apps save time, help officers focus on solving crimes — 33
Houston Police Department

Deploying an app to track fugitives — 37
San Juan County Sheriff

Adopting digital apps to secure the state fair — 42
Illinois State Police

Planning and managing security for a train celebration	48
Ogden City Police Department	
Securing the World Series with real-time technology	54
Cobb County, Georgia	

PART 3: CITIZEN ENGAGEMENT — 63

Using maps and apps to strengthen community-based policing	65
Toronto Police Service	
Mapping Canada's missing children to quickly reunite them with family	70
Missing Children Society of Canada	
Apps help provide targeted assistance to people experiencing homelessness	78
San Bernardino County Sheriff's Department	
Spatial analysis of opioid use gets lifesaving medicine to the right places	82
University of Tennessee at Chattanooga	

NEXT STEPS — 89

Contributors — 97

INTRODUCTION

IT'S AN UNDERSTATEMENT TO SAY THAT POLICING IN today's world is challenging. Many agencies face budget constraints, widespread staffing shortages, and rising crime numbers. Law enforcement agencies must do more with less as they meet higher service demands to safeguard communities. To meet these challenges, police agencies must maximize their return on investment (ROI). In other words, police must use every available asset wisely and effectively.

Increasingly, one of the most effective tools for police is technology. Law enforcement technology includes many and varied hardware and software systems. These solutions include mobile technologies, cameras, sensors, computer-aided dispatch (CAD), and records management systems (RMS). The combined data, software, and analysis tools from these systems give police new and innovative ways to fight crime, increase officer safety, reduce call response times, and even help police better connect with citizens.

Despite these advances, many police agencies still struggle to get the most out of the technology they own. Much of the technology and data police need is in siloed or single-use legacy systems. These legacy systems often are difficult to use and incompatible with new, modern technology and infrastructure. And although these sensors and systems collect data, typically they are not connected or accessible for decision-making.

Police agencies often encounter these common challenges:

- Legacy systems are hard to access and lack security.
- The lack of integrated systems creates data silos.
- Limited analytical capabilities lead to ineffective decision-making.
- The inability to share and collaborate with internal and external stakeholders leaves communication gaps.

Increasingly, police use geographic information systems (GIS) to address these and other challenges. GIS can serve as the foundation to integrate all the various systems, databases, and data types that every agency possesses. As such, GIS can rapidly process, analyze, and disseminate actionable intelligence. GIS can dramatically improve many common police workflows, including strategies to reduce crime and improve field operations. GIS integrates with many of the world's leading law enforcement solutions and can easily adapt to meet specific agency needs and workflows. This capability means that as agency needs change, the technology evolves with configurable tools to meet today's needs and prepare for what may come tomorrow.

As police departments analyze crime patterns, manage patrol operations, and support community engagement, the GIS toolbox helps manage, analyze, visualize, and share data via interactive maps and apps. Every day, police agencies worldwide use this technology to understand complex problems, increase decision-making capabilities, improve situational awareness, and collaborate to solve tough, real-world issues.

Analysis

Crime and intelligence analysts have used spatial analysis for decades to help police agencies detect, understand, and respond to crime problems. Traditionally, using spatial analysis has been a laborious process that required extracting data manually from an RMS, possibly processing and performing data analysis in older GIS programs, and exporting or printing static maps that weren't editable or secure. Analysis with today's modern mapping and spatial analysis tools looks much different. Analysts now use GIS to connect directly to RMS databases, automate the process, perform advanced spatial analysis in minutes, and then deliver interactive maps and apps for the rest of the agency. The work of analysts is foundational in helping law enforcement agencies make smarter, better-informed decisions that keep our communities safe.

Operations

GIS improves police operations by supporting decision-making at the executive level and delivering actionable intelligence to officers in the field. Commanders use strategic dashboards to implement modern crime-control strategies and monitor the success of ongoing police initiatives. GIS technology also supports integration with sensor systems including CCTV, remotely piloted aerial vehicles (UAV), body-worn cameras, license-plate readers (LPR), automated vehicle location (AVL), and other technologies that collect location information. Operation centers integrate these data sources into dashboards to support real-time situational awareness. Mobile apps provide additional operational support with tools for operations personnel, including navigation, field data collection, form surveys, passive location monitoring, and peer-to-peer communications.

Citizen engagement

GIS also provides new ways for police agencies to communicate and collaborate with the community. The community can access maps and apps via GIS citizen engagement hubs that encourage feedback and involvement in problem-solving initiatives. This technology can also promote transparency and accountability by sharing open data about ongoing and historic crime problems, police use-of-force incidents, and agency demographics. Many agencies use this same technology to support those experiencing homelessness, respond to current health crises such as the opioid epidemic, and combat human trafficking. The technology drives other collaborative initiatives by providing tools to collect, share, and collaborate with all stakeholders.

Stories and strategies

Protecting the People: GIS for Law Enforcement presents a collection of real-life stories that illustrate how organizations use GIS to enable data-driven crime-fighting strategies, drive decision-making and operational awareness, and promote community policing initiatives. The stories and strategies present The Geographic Approach to integrating GIS and spatial reasoning into law enforcement operations. The book concludes with a section about next steps with GIS, which provides ideas, strategies, tools, and suggested actions organizations can take to build location intelligence into decision-making and operational workflows.

This book presents location intelligence as another crucial layer of knowledge that managers and practitioners can add to their existing experience and expertise. If location intelligence isn't already part of an organization's decision-making processes, considered in daily operational activities, or used to improve constituent satisfaction, managers can use this book to start developing skills in those areas. Developing these skills does not require GIS expertise, nor does it require managers to disregard their experience and knowledge. The Geographic Approach adds another way to think about solving problems in a real-world context.

HOW TO USE THIS BOOK

THE GEOGRAPHIC APPROACH IS USED FOR LOCATION-BASED analysis and decision-making. GIS professionals typically employ it for the comprehensive study and analysis of spatial problems. This book is designed as a guide to help you take first steps with GIS to address issues that are important to you right now. It will help you apply The Geographic Approach to decisions and operational processes for solving common problems and creating a more collaborative environment in your organization and community. You can use this book to identify where maps, spatial analysis, and GIS apps might be helpful in your work and then, as next steps, learn more about those resources.

You can also learn about additional GIS resources for law enforcement by visiting the web page for this book:

go.esri.com/ptp-resources

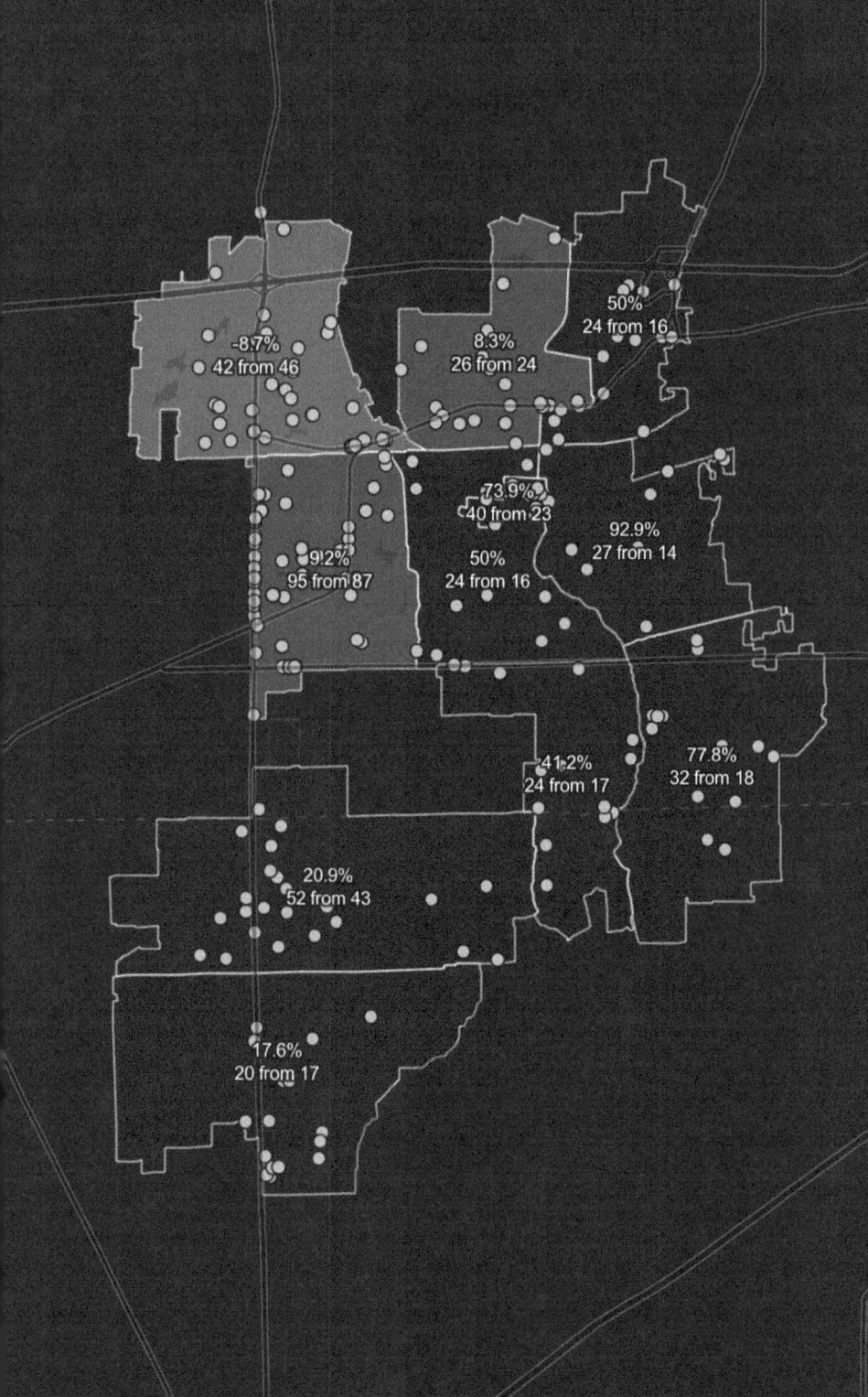

PART 1

ANALYSIS

SUCCESSFUL DATA-DRIVEN POLICING STRATEGY STARTS with good analysis. Law enforcement agencies depend on the work of crime and intelligence analysts to reduce crime, support investigations, improve operations, and make smarter, data-driven decisions. Crime and intelligence analysts use modern GIS mapping and spatial analysis tools to make sense of large amounts of data and deliver analysis results to officers and commanders to make better-informed decisions. Because GIS is easier than ever to use, analysts spend less time preparing data and more time enabling decision-making.

Manage

To do their work, analysts must access data from agency databases and information systems that house incident, crime, offender, and sensor information, and other types of data. A GIS can act as a system of record for these disparate data sources and help an analyst extract, integrate, and prepare data for analysis. Automating these processes can make data readily available for analysts to spend less time preparing data and more time performing analysis.

Analyze

Many of the most common types of analysis are spatial, and connecting people, events, and places together temporally and spatially is the basis for solving many types of crimes and criminal associations. Analysts use GIS to map incidents and identify series, patterns, trends, and hot spots of incidents in support of short- and long-term crime control strategies and to aid investigations by identifying and linking criminal networks and activities. Today's modern GIS technology can do even more, as analysts use advanced techniques such as spatial statistics, machine learning, and 3D models for even deeper understanding of complex patterns and relationships in the data.

Share

Analysts must get information products out to the rest of their agency. GIS allows them to share analysis quickly and easily using maps, data visualizations, charts, and reports delivered in a variety of formats, including interactive bulletins, web maps, mobile apps, and hard-copy printed documents. From there, the logical next step is providing maps and apps for the end user to do their own analysis and query the data. Creating dashboards for command staff, mobile apps for officers in the field, and public web apps to share information with the community are relatively easy tasks using configurable GIS web apps.

GIS in action

This section will look at real-life stories about how law enforcement organizations use GIS analysis to gain better insights.

SMART POLICING GETS A BOOST FROM ENTERPRISE GIS

St. Petersburg Police Department

For years, the police department in St. Petersburg, Florida, relied on paper and spreadsheets to fight crime.

"Everything was text based," said Frank Ullven, a systems analyst on the St. Petersburg Police Department's Information and Technology Services (ITS) team. "We didn't have any maps. It was all street names and addresses."

Ullven remembers how police officers in the Gulf Coast community had to read addresses in columns to figure out where incidents were occurring. Matching a crime to an address was difficult without visual representation, especially when new streets were added to the community or street names were changed.

"Officers don't have to memorize all that, like this used to be 2nd Street South, but now it's University Way South," said Ullven.

St. Petersburg Police Chief Anthony Holloway advocated a policing method called CompStat, which uses timely and accurate information to combat crime effectively and improve police accountability.

The police department's ITS team, which includes a GIS specialist, systems analyst, and data specialist, along with a team of five analysts in the Intelligence-Led Policing (ILP) unit, set out to address and fix the lack of visual representation. Together, the group administers ArcGIS® technology and creates the maps and dashboards used by more than 575 officers, detectives, and supervisors at the police department. More specifically, the ITS unit manages the department's ArcGIS Enterprise portal, while the ILP unit is a data-driven center that provides support to tactical, strategic, and operational initiatives.

According to Dr. Richard Ferner, Jr., the ILP unit supervisor, several department stakeholders, including the police chief and command staff, were overwhelmed by the sheer volume of text-based information about crime. The police department implemented older types of geospatial technology but lacked the flexibility to support custom, user-friendly visualizations.

"Those solutions did not promote robust and relevant visualizations," said Ferner. "Users had little incentive to utilize those tools when deliberating on a course of action, such as proactive patrol assignments or developing leads in identifying suspects." Because of inadequate visualizations, staff and supervisors had difficulty gaining meaningful insights to make decisions and work effectively.

In 2014, the arrival of the new police chief, Anthony Holloway, marked the department's transition to a data-driven organization. Holloway advocated adopting a management model called CompStat, or computer statistics, a policing method that uses timely and accurate information to combat crime efficiently and improve police accountability.

At the time, the ITS team realized it needed to move away from a static environment and deliver content in an interactive manner.

"The minute I heard Esri® developed an enterprise solution that could allow the user community to interact with the content we publish, I knew, unequivocally, that was the solution we needed," said Ferner.

In 2016, the department implemented ArcGIS Enterprise and ArcGIS Pro software and has kept pace with each upgrade, steadily adding products such as ArcGIS Insights[SM] and ArcGIS Dashboards.

ArcGIS Enterprise was key in supporting the department's need for a secure, behind-the-firewall enterprise platform that powered data management and analysis, especially in the context of law enforcement data.

"There's a level of comfort in knowing that it's our data on our in-house system and that it's not located somewhere that we don't have control over who sees it or what's being accessed," said Ullven.

Before implementing ArcGIS Enterprise, department analysts created static content that was distributed through email and posted to a file-sharing system and Microsoft SharePoint. Disparate tables, charts, and graphs did not tell the whole story, and there was no way to customize dashboards to visualize data and come up with a common operating picture.

Using ArcGIS technology, "we can carve out highly nuanced, relevant data that matters and answers questions," said Ferner. "It helps the staff and supervisors carve out a strategy and set of tactics for immediate application."

Analysts have created dashboards and stories that focus specifically on what each unit needs to know. That way, people don't get overwhelmed with irrelevant information.

The department also began refreshing the data 45 minutes before each shift, allowing watch commanders to detect emerging crime trends and evaluate initiatives on the go. Analysts use the Crime Analysis configuration in ArcGIS Pro along with Insights to analyze data and then share interactive content via stories and dashboards made using ArcGIS Enterprise.

"Canvassing a neighborhood no longer requires a six-foot-long paper map and tons of hard-to-decipher markings," said Kevin Christy, the ITS team's GIS specialist.

Instead, an app created using ArcGIS Web AppBuilder made the process more targeted and efficient by allowing detectives to digitally track the addresses they visit. And ArcGIS Survey123 enhanced the department's Eagle Eye program, a public camera registration website, making it easier to geocode addresses and maintain an up-to-date camera locator app.

In one example, the command team was looking for information about parking meters being destroyed in downtown St. Petersburg. Analysts pulled crime data for areas around parking meters, used ArcGIS Pro to predict a crime risk area, and published this data on a map in ArcGIS Enterprise. Staff could then use the prediction to plan operations.

The data to make the prediction was acquired from the records management system (RMS), where detailed accounts of crime around each parking meter location were documented. Analysts

With ArcGIS, analysts can create dashboards and stories that focus on what each unit needs to know to fight crime in the area.

then geocoded each parking meter location in ArcGIS Pro and subsequently published the content as a hosted feature layer in ArcGIS Enterprise.

Armed with this analysis, police officers patrolled the risk area identified in the prediction and encountered the suspects, who were arrested as they prepared to commit more crimes. Prior to implementing this vast array of ArcGIS technology, doing this kind of analysis and making such a prediction were not possible.

Now, more officers are requesting specific dashboards from the ILP and ITS units. They want to see what ArcGIS technology can do, and when they get a tour of it, their eyes light up, according to Christy.

"I see their wheels turning," he said. "The big thing is tailoring it for exactly what the end user wants. Whether it's 'I want these metrics in my dashboards' or 'I want these colors' or 'I want an app that does x, y, z,' it's all about giving them what they want. If they don't get exactly what they want, they're less likely to use it."

Ferner also noted that having a growing number of younger police officers has contributed to a critical mass of software users in the department.

"One of our biggest challenges was the cultural dynamic in giving them access to these products and assigning accountability to the metrics," he said. "The workforce here is also becoming younger, and we've discovered that they are more adept with using different technologies. Even if we had a product 10 years ago that is as sophisticated as this one is today, I don't think the staff then would have been so accepting of these technologies."

The move from static maps and data that officers couldn't fully engage with to a more interactive mapping platform has transformed the department.

"Esri allows officers to have a customized product that really presents them with geospatial data that prompts questions," said Christy. "They can look at the data, they can ask questions, and now they are getting more insights than they had in the past. It's really helping to drive better policing."

A version of this story originally appeared in the spring 2020 issue of *ArcNews*.

COLLABORATING TO ADDRESS CRIME THAT ARRESTS ALONE WON'T SOLVE

Philadelphia Police Department

LIKE MUCH OF THE COUNTRY, PHILADELPHIA HAS BATTLED an opioid crisis of unprecedented proportions, with residents suffering from overdoses in record numbers. The epidemic has surged in certain hot spots, notably the city's Kensington neighborhood.

"Kensington is ground zero for the opioid problem on the East Coast, and one of the country's biggest heroin markets," said Kevin Thomas, director of research and analysis for the Philadelphia Police Department (PPD).

Law enforcement officers in Philadelphia confront the problem by doing more than just arresting people for drug offenses. PPD uses The Geographic Approach—through a GIS that enables analytics and collaboration with other city agencies—as a key part of its response to the ongoing public health emergency.

The team's analysis takes several forms, including correlating the locations of overdoses. For example, most heroin sold on the street is identified by its "heroin stamp," in other words, the images and icons dealers emboss on the bags they sell. PPD officers can trace the stamps responsible for geographic clusters of multiple overdoses and connect them to the originating street corner and drug organizations.

"We can track overdoses, so we have that data all in one place," Thomas said. "We also built an application to track the stamps. Combining those two data-gathering efforts and the intelligence that supports them, we can say, 'Okay, these are the three corners that you need to send your police officers to shut down narcotics operations, because we want to tamp down that overdose spike.'"

Using GIS technology allows police to focus on the geographic

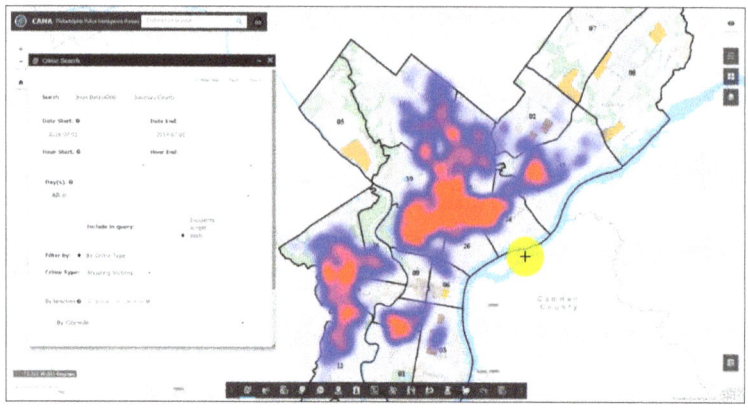

Looking at the locations of crime across a time span allows Philadelphia police to spot patterns and understand if their practices make a difference.

locations where the epidemic is most concentrated. In this way, a seemingly intractable problem that can't be solved merely by arrests becomes more manageable.

This work relates to a larger PPD initiative, Operation Pinpoint, launched in 2019. The operation aims to combine data-driven information from various sources with real-time, ground-level intelligence from officers on patrol. Pinpoint enforcement zones are selected based on data analysis. Officers use this analysis to plan when and where their outreach will have the most impact.

Determining optimal deployment

The GIS aspect of Operation Pinpoint has allowed for a useful feedback loop, as information regarding the location and activity of officers shapes future analysis. Every five seconds, a mobile data computer in each PPD vehicle calculates and transmits the vehicle's position.

"We can present information to command staff, supervisors, and commissioners to help them evaluate and adjust their deployment,"

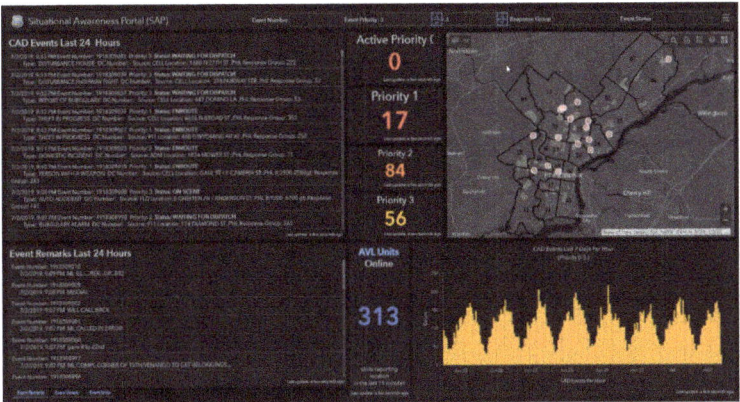

A dashboard view presents daily incidents and officer actions to highlight existing and emerging hot spots.

Thomas said. "It's about understanding deployment patterns using a passive measure like GPS [location], instead of, say, being so focused on things like stops, arrests, and curfews. We're trying to integrate that back into CompStat so that we can understand how much proactive police time is going into certain grids."

CompStat is a catch-all term for a type of approach to law enforcement, driven by statistics and pioneered in New York City in the 1990s. The approach is now considered a standard accountability practice in many police departments, including Philadelphia's. GIS has always been integral to the CompStat approach, and using GPS data represented a logical next step. As the amount and diversity of data available to police increases, GIS provides a common platform and framework for warehousing data and providing analytics. Because it displays data in the context of physical location, GIS allows police to visualize the complex workings of the city with more clarity.

In Philadelphia, the police themselves can become an important data point. The presence of cluster points where police cars coalesce can highlight new hot spots. "The saying in CompStat," Thomas said, "is what gets measured gets done."

Looking at problems in a new light

Location data helps PPD design unique policing plans for different areas of the city, especially the 46 designated hot spot regions where violent crime is most prevalent. "Each one of those targeted areas should have an articulated plan, so that everyone, all different aspects of the organization, can understand what their role is in supporting the solution," Thomas said. For Kensington, locus of the opioid trade, police encounter a hot spot area with people coming and going and a constantly changing set of offenders.

"What we're trying to do is look at things on a neighborhood basis, not a topic basis," Thomas said of the GIS-based approach, which allows PPD to see how local characteristics can be a cause and symptom of crime. "Where you have violence, you'll also have blight problems. You'll see income and poverty issues. And you'll see opioids. Having that location focus is perfect for our GIS platform."

The GIS platform for the PPD combines information driven by intelligence—such as observations made by police—with data-driven metrics.

"That combination can inform decisions regarding, for example, where a task force should be or where we should focus on offenders," Thomas said.

This approach can also integrate data from other city agencies, leading to a more holistic approach to policing.

"Opioids and the violence caused by the drug trade are not just police problems," Thomas said. "They're city problems. Solving them requires the input of other city agencies, quasigovernmental agencies, and even the private sector."

The use of GIS to visualize problems in the context of location

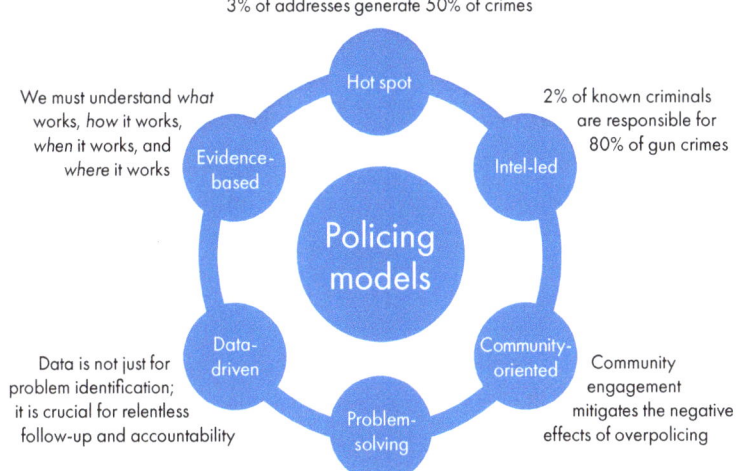

Police use various data-driven models in their work to combat crime, and specifically, opioid use, in Philadelphia.

along with collaboration across city and private agencies are part of a uniquely ambitious plan to revolutionize the process of urban law enforcement.

"This is not a short-term solution, something we're using to just knock down crime this summer—we are changing how the organization operates," Thomas said. "It's changing how we communicate, how we do police work, how we share information and intelligence, and how we coordinate goals and objectives. This is about much more than just crime reduction."

Transitioning to a cloud platform

A project like Operation Pinpoint involves a staggering amount of data. Any method of storing and processing this information must balance privacy concerns with flexibility. It must also comply with the strict protocols of the FBI's Criminal Justice Information Services Division (CJIS).

For the PPD, the natural choice was a cloud-based system. Building Pinpoint around Microsoft's CJIS-compliant Azure Government platform eliminated the need for 25 on-premises servers. It also simplified the process of importing terabytes of data into a GIS environment.

"In terms of storage and computing capacity, the cloud is phenomenal," said Andrew Smart, senior GIS consultant at geographIT, a division of EBA Engineering, an Esri business partner helping PPD establish and manage its Azure cloud platform that makes Pinpoint possible. "We can quickly provision servers to support the platform without having to acquire and wire physical hardware as well as continually monitor and scale the environment to support the growing needs."

PPD's Kevin Thomas agreed. "We went to the cloud because we lack the necessary amount of internal IT staff," he said. "Due to slow procurement processes and lack of IT staff, it is incredibly difficult for us to maintain servers on premises and refresh them every five years."

The ability of Azure to help PPD scale the project up or down as needed was apparent from Pinpoint's earliest implementation. Soon after the launch of the system that displayed the real-time location of PPD's fleet, Smart placed an urgent call to Thomas. The Automatic Vehicle Location (AVL) server was operating over capacity, dropping data as it attempted to process the overload.

It turned out that the system, which was supposed to roll out gradually to devices, had been pushed to every car at the outset. The deluge of 400-plus records per second from more than 500 cars was too much for the server to handle in its current configuration.

"If this was an on-premises box, we would've been in trouble," Thomas said. "But we could just go to that server, increase the resource in Azure, and stabilize the environment within five minutes from the time we received an alert about the issue. All of a sudden, you could see usage drop to a reasonable level."

A version of this story by John Beck titled "Philadelphia Police Collaborate to Address Crime that Arrests Alone Won't Solve" originally appeared on the *Esri Blog* on August 28, 2019.

SHARING MAPS, APPS, AND DASHBOARDS TO IMPROVE PUBLIC SAFETY

St. Louis Metropolitan Police Department

AT TIMES, THE CRIMES THAT PEOPLE COMMIT ARE SO brazen, so out in the open.

"They don't seem afraid of getting caught or of the consequences," said Captain Christi Marks, a commander with the St. Louis Metropolitan Police Department (SLMPD).

In response to bold retail crimes in St. Louis, Marks and her team took The Geographic Approach—using maps and dashboards to locate crime patterns and allocate resources where they are needed most.

Mapping crime to find trends and protect residents

Marks had long used maps to understand crime trends in the city, beginning with a paper map and different-colored pushpins she used to mark each crime by category. Marks used this approach in the early 2000s when she was a public affairs officer for the downtown district.

"As you can imagine, by the end of the year that map had thousands of holes in it," she said. "I told my captain, 'There has to be a better way.'" After investigating crime analysis and visualization tools, she took GIS training to compile data and make digital maps.

"I remember how proud I was when I bound up all these crime maps in booklets," Marks said. "Other captains were saying, 'Where did you get that?'"

Since 2007, St. Louis police crime analysts have used GIS daily to inform patrol plans, aid investigations, and help allocate resources. The agency expanded access to GIS at the enterprise scale, giving the force of 2,000 people access to location intelligence.

PART 1: ANALYSIS   17

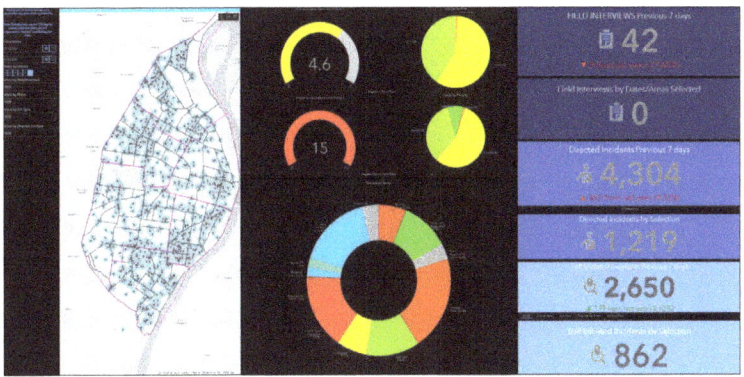

SLMPD's CAD dashboard shares details on crimes and calls for service.

Emily Blackburn, the Crime Analysis Unit manager, worked with crime and intelligence analyst Lindsay Maier to create interactive data dashboards with built-in filters that anyone on the force can use to focus on specific locations or trends.

Blackburn and other crime analysts and researchers have learned that crime is highly concentrated, with 5 percent of the locations in a city accounting for 50 percent of that city's crime. They also know that repeat offenders commit the most crimes. These patterns inform decisions to increase police presence where it's needed most.

One of the GIS tools Blackburn and her team created is an application that combines data such as details from field interviews about crimes and information about the people who may have been involved. Officers can search the app when working to catch suspects and solve open investigations.

"We can look up any suspect's street name in St. Louis, and you can use that or any alias in the app to get all frequented locations," Blackburn said. "It really creates a good picture of a suspect, where they might be, and if they might be the person they're looking for."

Another app created for police commanders shows a side-by-side map of crimes and police activity. "Now they can really see almost in real time if officers are where the crime is," Blackburn said.

Data-driven public safety during the pandemic and beyond

The pandemic brought new challenges to the crime analysis team but also new opportunities to socialize their work and offer support for COVID-related shifts in crime patterns. The team also used the time to improve its data consolidation and automation.

"It gave us back our commute time, so we had time to think creatively and do more with our data," Blackburn said. "We also pulled back from police-initiated activity until we learned how to manage COVID-19." The shift gave the department time to focus on taking GIS to every officer.

The rising popularity of public health dashboards increased awareness and adoption for police officers. Blackburn explained: "When I shared the dashboards, everybody said, 'Oh, this is like the COVID-19 dashboard,' and I said, "Yes, except for crime.""

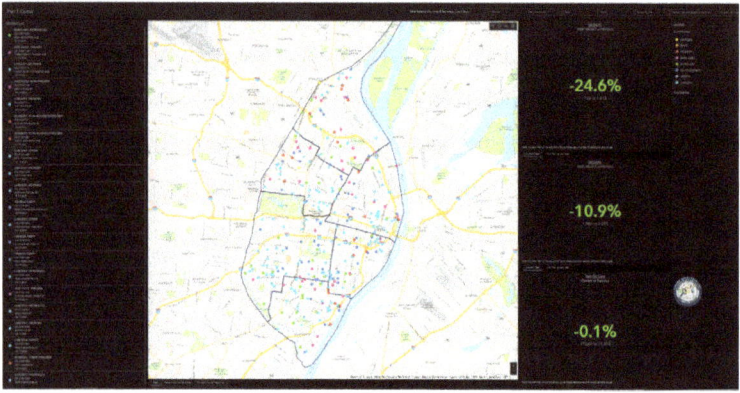

SLMPD takes a CompStat approach to crime, using computers and dashboards to track and monitor performance.

The police dashboards quickly oriented officers on how COVID-19 was impacting the areas they patrol. Maps show, for example, the number and locations of domestic violence incidents and suspicious deaths and overdoses—all of which became even more concerning during the pandemic.

"I get on the dashboards every morning for the daily stand-up meeting with all the commanders, captains, majors, and colonels—Monday through Saturday," Marks said. "I can never be blindsided because every single incident that's reported pops up, and it also tells me what's up or down. I know exactly where everything happened, and I can click on just the dot to learn more."

As US cities work to recover from the pandemic, Marks hopes the added awareness from GIS apps and dashboards will keep residents safe and help return her city to vitality.

"You can go downtown on a night where there's a baseball game, and you'll see thousands and thousands of people," Marks said. "And when I see that, it makes me so proud, because that's my city and that's what I want to see. I want to see the commerce. I want to see people coming to downtown. I want a thriving city."

A version of this story by John Beck titled "St. Louis Police Share Maps, Apps, and Dashboards to Improve Public Safety" originally appeared on the *Esri Blog* on October 26, 2021.

ENHANCING STRATEGIC PLANNING AND CRIME ANALYSIS WITH DATA ANALYTICS

Altoona Police Department

LAW ENFORCEMENT OFFICERS IN ALTOONA, A SUBURB OF Des Moines, Iowa, serve and protect a community of 20,000 residents. Add in visitors, and that number can increase by thousands. With a shared vision of making Altoona the place of choice to live, work, and play in Iowa, the Altoona Police Department combines traditional policing with new technology to prevent and solve crimes and keep residents safe.

Data-driven policing involves collecting and analyzing data to help target specific areas, better allocate resources, and make more data-driven decisions to protect residents. To support this approach, Altoona police started an initiative to strengthen strategic planning and reduce crime by using all relevant data, including location-based data. The department implemented software for spatial and nonspatial data analysis. This implementation helped police more effectively allocate resources and personnel and conduct crime analysis.

Challenge

Altoona police uses CompStat, an organizational management system for police departments that combines management techniques, crime analysis, and mapping technology. However, the lack of timely data and generation of reports made quick data-driven decisions a challenge.

The process previously in place meant that Tony Chambers, a captain with Altoona PD, had to extract data through Structured Query Language (SQL) for storing, manipulating, and retrieving data in databases. Chambers would place the data in spreadsheets, use pivot tables for analysis, and finally produce static reports.

However, to support the department's efforts to understand what and where things were happening in Altoona, Chambers wanted to infuse more location-based data with CompStat to better allocate resources.

This allocation involves tasks such as analyzing data to determine officer workloads and using data to justify the need for additional resources.

"It's difficult to allocate resources to areas unless we're able to map this stuff. It's great if I just create a bunch of reports with stats on them, but as far as operations, decision-making, or resource allocation, you still don't know the 'where,'" said Chambers. "We needed to tie all of that information that I was putting into those static reports to locations."

The department also needed more timely information. According to Chambers, the data was already several weeks old when the department produced and distributed its reports because of the time it took to compile data in CompStat and put the data into a pivot table for analysis.

In addition to improved operational decision-making, Chambers wanted a tool that police department staff could easily access and use to present information clearly.

"Staff come and talk to me about this data and want to know more about it. I want to be able to put this information—the data, the analytics—at the fingertips of even our patrol commanders and sergeants," said Chambers.

Solution

Chambers set out to find a solution to collect and analyze data for operational decision-making, resource allocation, and workload analyses to strengthen strategic planning. He was already a user of ArcGIS Pro, a desktop application for data analysis and 2D and 3D map creation, and ArcGIS Online, the cloud-based GIS mapping software. So Chambers decided to explore other Esri offerings for an especially smooth transition.

Chambers attended a local workshop on ArcGIS Insights and recognized its potential benefits for the department. Using a free trial, he explored the software's capabilities for analyzing spatial and nonspatial data, fusing location analytics with open data science.

Insights seamlessly integrated with the police department's existing ArcGIS Online platform. The web-based software made installation easy, and the department had the options of using Insights on the desktop and with ArcGIS Enterprise.

"I was looking for a system or program that we could put into place that anyone can use, that doesn't require you to have programmer skillsets or be an analyst," Chambers said. "ArcGIS Insights fit the bill. [Police officers] should be able to access data in a way that

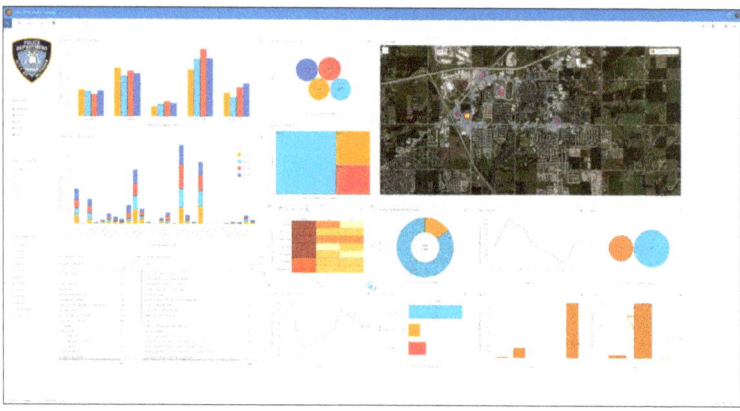

Annual report of arrests during a four-year period for comparison. Each metric on the Insights page is selectable to filter the results based on the selection. This page includes Uniform Crime Reporting (UCR) offense type, category, and charges, by year, day of week, time of day, type of arrests, and demographics. The heat map in the upper right shows the location of arrests.

they can easily consume and digest it to know what's going on in their districts."

Chambers used Insights to perform different functionalities in one place.

The key moment for Chambers came when he realized he could use location analytics while also building charts and data tables and doing everything he needed using pivot tables in a spreadsheet program.

Results

The use of Insights allowed Chambers to produce dynamic reports and provided senior officers the information they needed to make better decisions. Insights allowed police officers to interactively explore data to understand what was going on in their districts and solve problems on their own.

Insights also allowed first-line supervisors and commanders to view and access data faster, an improvement from the old method of compiling data in spreadsheets and making it available only during monthly meetings. Using Insights, Chambers produced dynamic reports as opposed to static ones, which let users engage with data and details.

"Instead of us getting together to view CompStat and identifying an area we need to look at, I expect that the district lieutenant already knows about what's going on in that area and is already working on a solution for the problem. The whole process becomes more efficient and fluid," said Chambers.

Another significant benefit of Insights is the ability to uncover patterns and discover relationships in the data. Previously, Chambers would load data into a spreadsheet, update the pivot tables and pivot charts, paste their pictures into a Word document, and publish

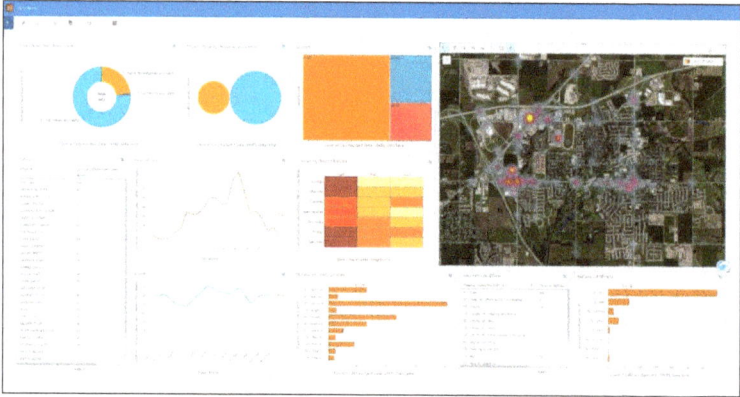

This dashboard shows the statistics of traffic accidents during a year's time to find trends by month, day of week, time of day, districts, and area. Commanders can use this map to analyze a particular roadway or intersections and determine the type of accidents occurring in those areas.

it. But it can be difficult to connect and find relationships from static reports and maps.

"Data-driven policing is just a relentless analysis of data. So for us to be effective, we need to constantly reevaluate the statistical information along with the location information. And ArcGIS Insights allows us to do that much more efficiently and effectively," said Chambers.

It's also more efficient to edit maps, charts, and other visuals created in Insights than it was using the previous method, Chambers said. He no longer must import data into several different spreadsheets and then refresh them each time. Insights lets him update the dataset in one place, and Insights updates charts, tables, and maps.

Chambers looks forward to using more of the Insights analytical tools. The data analysis and results have helped him with everything from making daily operational decisions to creating more strategic planning-type reports, all in one place.

"Esri is constantly making improvements, which gives us more functionality and features. So it really seems like the possibilities with Insights are endless," said Chambers.

A version of this story titled "Altoona Police Enhance Strategic Planning and Crime Analysis with Data Analytics" originally appeared on esri.com in 2020.

PART 2

OPERATIONS

POLICE OPERATIONS ENCOMPASS THE ESSENTIAL activities that police departments engage in every day to keep our communities safe. These include responding to calls for service, engaging in directed patrols, and preventing and investigating crimes. But operations can include other activities such as managing special events and responding to disasters or extremely dangerous situations, such as an active shooter. GIS can help improve police operations at every stage with tools for mission planning, sharing and implementing plans, tracking and monitoring ongoing missions, and capturing and reviewing key metrics of completed operations.

Plan

Operational planning starts when an analyst finishes and shares their work with the operations supervisor. After reviewing the analysis, a supervisor better understands problem locations and knows when and where to deploy resources for greatest effect. Plans created interactively with maps and apps allow the supervisor to see hot spots of criminal activity, locations of targets, special event requests, and any other information or analysis that warrants an action plan. Typically, the planning app will be a web application that can be marked up with activity zones, officer posts, event routes, rally points, and whatever else is specific to that plan.

Operationalize

Once a mission plan is ready, field supervisors can brief other officers and other personnel. Patrol officers can view the plan on in-car computers and mobile devices and see their current assignments, find and navigate to locations, collect and edit information, and share their location and status in real time. Line supervisors, command staff, and operations centers typically have access to key operational information. Police can monitor all activities in real time using dashboards that integrate live data feeds, track personnel locations, and allow staff to push updates and information to the field as the situation changes.

Review

Commanders must review ongoing and completed operations to understand what worked and what didn't. They can use GIS tools to analyze, evaluate, and measure key performance indicators such as crime numbers, arrests, case clearance, and distribution of officers geographically. Using GIS dashboards, police can review maps, charts, graphs, and reports to track metrics, achieve objectives, and improve community service.

GIS in action

This section will look at real-life stories about how law enforcement organizations use GIS to improve operations.

SAVING TIME AND MONEY BY INCREASING FIELD DATA COLLECTION EFFICIENCY

Redlands Police Department

CALIFORNIA STATE LAW REQUIRES EVERY SEX OFFENDER living in the state to register as such with the police department of the city or county where they reside. Offenders must register anew every year or whenever they move. To ensure that Redlands registered offenders comply with California Penal Code 290, the Redlands Police Department (RPD) makes unannounced visits to each offender's home annually to verify residency and compliance with parole terms.

In the past, the Redlands police began their annual checks by printing out copies of each registrant's data and contact information. Teams of RPD officers would take these copies and complete field interviews to verify compliance. Using pen and paper, officers would cross names off the list as they finished the checks. Back in the station, officers updated information and manually entered it into spreadsheets.

Challenge

This method of planning and managing these checks was labor-intensive and time-consuming. Staff would typically take weeks to plan and days to complete the home checks of approximately 125 sex offenders. Sharing the data manually caused delays. Up-to-date information wasn't available until after officers in the field had completed their interviews and the data was entered into the RMS.

Solution

RPD recognized that the manual process was antiquated and wanted to find a way to modernize workflows. To do this, the department needed to streamline the planning and execution of the operation, get a better idea of what is happening around them in real time, and improve the sharing of information after each operation. RPD turned to digital reporting for increased efficiency in the field and live monitoring in the command center. The department realized it could use its existing GIS for better planning and improved collection and sharing of data.

Using Esri technology, the RPD digitally mapped the locations of its sex offenders and shared the information with officers in the field via the Survey123 mobile application. The department then configured ArcGIS Dashboards in the command center to monitor their operations and officer locations in real time.

After mapping each of the offenders' registered addresses, police created geographic zones and assigned them to RPD teams. This process increased efficiency, which allowed teams to focus on offender locations based on proximity, reducing unnecessary travel between visits.

As officers collected data in the field, staff updated the maps in real time in the command center.

Digitally mapping the locations of offenders enabled the RPD to share the data with officers in the field via the ArcGIS Survey123 mobile app.

Using Survey123, the officer in charge shared the data with the task force of 24 officers within minutes. The task force used the app to collect data during each residence check, update the information, and take a photo to include in the database.

As officers collected data in the field, maps in the command center were updated in real time. The dashboard allowed the police chief and command staff to monitor each team's location and track completed interviews as they were updated. This information supported real-time decision-making as the operation progressed.

Results

Officers using Survey123 completed their checks in significantly less time than the old labor-intensive pen-and-paper method. As officers updated records in real time, records across the system updated instantly. This method streamlined workflows and maximized operational efficiency.

With geocoding and analysis done ahead of time, the route for each team was planned for maximum efficiency to minimize redundancy. Mapping each of the sex offenders' residences and grouping them into zones in ArcGIS Online helped the RPD save time and mileage.

The success of this operation and the immediate returns on investment led the department to geoenable additional police operations. Using maps and apps, the RPD hopes to improve crime reduction strategies, support other police initiatives, and find innovative ways to use technology for a safer Redlands.

A version of this story titled "Police Department Saves Time and Money by Increasing Field Data Collection Efficiency" originally appeared on esri.com in 2019.

PURPOSE-DRIVEN APPS SAVE TIME, HELP OFFICERS FOCUS ON SOLVING CRIMES

Houston Police Department

POLICE HAVE LONG USED GIS TECHNOLOGY TO HELP identify crime patterns and decide where and when to assign resources. But modern GIS technology can do much more by giving police the ability to create an enterprise environment that supports every mission, including advanced analytics, dashboards to fit many needs, public engagement apps, and maps and apps that provide real-time situational awareness. The Houston Police Department (HPD) is one of these agencies, using enterprise GIS to save time, streamline operations, and better serve its city.

One of HPD's latest initiatives is to share data and analysis with officers as quickly and efficiently as possible. The HPD Crime Suppression Teams (CST) are dedicated tactical units that consist of one sergeant and eight officers who use analysis to identify crime trends and patterns and initiate proactive policing methods. To get analytical products in officers' hands more efficiently, the department's GIS Unit was tasked with creating a mobile app to provide the most current information and promote situational awareness for all team members.

As with many other cities, Houston began 2020 with a vibrant economy. But as COVID-19 continued to spread and stay-at-home orders were issued, businesses shut down, and unemployment claims rose. Many American cities, Houston included, experienced increases in crime, including a year-over-year increase in gun violence and homicides. HPD also began seeing rising incidents of street robberies and burglaries associated with two specific methods of identifying and targeting victims. These methods are known as "jugging"

Houston PD officers are able to analyze patterns and hot spots of incidents in the field.

and "sliding." In jugging, offenders watch locations such as ATMs or bank drive-throughs to target people who are carrying money. The offender then follows the victim to another location to rob them or burglarize their car or residence. Tracking the incidence of jugging can be difficult because victims often have no idea how they were targeted or followed.

In sliding, offenders target victims while they are stopped at gas stations and are preoccupied paying and filling their tanks at fuel pumps. The modus operandi of these kinds of thefts usually involves two or more perpetrators. The suspect vehicle, often with tinted windows to conceal the backseat passenger, pulls next to the victim's vehicle, and the perpetrators then quickly enter or "slide" into the targeted car and steal whatever they can grab while the victim is distracted at the pump. The suspects then drive away before the victim even knows what has happened. As with jugging, sliding can be hard to identify, and incidents are often underreported or reported as other crime types.

Tasked with identifying and stopping jugging and sliding incidents, the CST team needed the help of the GIS Unit to build a mobile app. Working together, they created a dashboard that provides officers with an all-in-one self-service app that includes a map of current and historic crime incidents, the ability to filter them by incident type, and the ability to read reports in the app instead of filing time-consuming requests with a crime analyst.

"Staff wanted a way to do things out in the field such as capturing data on-site and taking photos," said Lt. Freddy Croft, GIS Unit supervisor. "The new app made data management much easier." Officers can use the app in their patrol cars, saving time on data gathering, organizing, and presentation by the analysts. Information that once took analysts two hours a day to produce is now available to officers instantly. Using the app, officers have the tools to answer

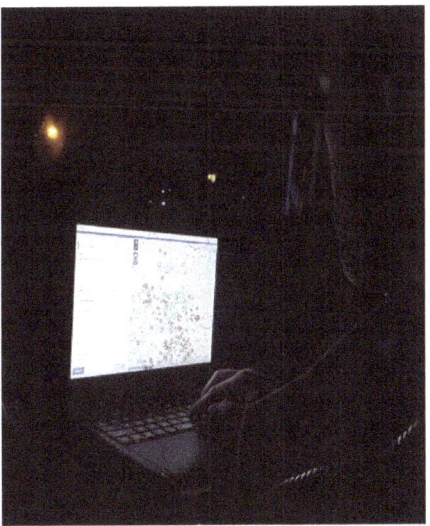

Understanding where incidents occur helps the HPD address crime surges and allocate resources efficiently.

"when and where" questions so they can focus on the "who" while conducting investigations.

The app provided analytics to nontechnical users, but a one-size-fits-all approach did not work for all the agency's needs. Using the same technology, the GIS Unit had the flexibility to build purpose-driven apps that fit the needs of current agency missions and the unforeseen missions of tomorrow. As new projects and requests come in, the GIS Unit responds by identifying the appropriate tools to fit new missions and the needs of end users. For example, the unit has created dashboards that show recent crime hot spots, current calls for service, and gang information. Officers can access this data immediately in their in-car computers and mobile devices.

As the GIS Unit gained experience in building these apps, it helped other units and teams identify the best GIS applications to fit specific business needs. Many of the ideas generated by the creation of the CST app made it possible to share information and best practices with other agencies and GIS users in the Houston community. The technology has the flexibility to fit any mission once the conversation's started.

A version of this story titled "Purpose-Driven Apps Save Time, Help Houston Officers Focus on Solving Crimes" appears in the book *Measuring Up: The Business Case for GIS, vol. 3* (Esri Press, 2022).

DEPLOYING AN APP TO TRACK FUGITIVES

San Juan County Sheriff

EVERY WEEKEND, THE POPULATION OF FARMINGTON, NEW Mexico, explodes. With a population of 120,000 people, the largest city in the mostly rural Four Corners region draws thousands of additional weekend visitors to shops, restaurants, movie theaters, and bars. Crowds create new challenges for police when weekday service calls more than double on the weekends.

Calls to keep the peace can stretch into the early hours when those who come to party cause trouble. Arrests often result in open warrants because defendants flee or fail to show up for court.

"It was hard for deputies to search and see open warrants," said Eli Lisko, information systems manager for the San Juan County Sheriff's Office and a reserve sheriff's deputy. "Deputies were typically going after warrants they obtained themselves as part of an investigation or looking for specific people from recent high-profile

The rugged terrain of San Juan County provides ample places for fugitives to hide.

cases. They didn't have the ability to effectively target older warrants or see where they needed to do house checks."

Now, deputies can use an app, powered by the county's GIS, to see all active warrants near their location. The app reveals a drop-down list of crimes for people with open warrants, and when a deputy chooses a warrant to follow up, the app routes them to last known addresses.

"Apps have to be practical, easy, and straight to the point for our deputies to even want to use them," said Sherice Snell, GIS manager for San Juan County. "We've had good success with this app because it makes the job much easier."

Tracking down fugitives

When there are no active service calls, patrol officers can use the app to pursue the thousands of active warrants in San Juan County.

"This provides deputies with another opportunity to be proactive in the community instead of just finding busywork," Lisko said.

The job of finding people with active warrants comes with many challenges, including the knowledge that some will fight to stay out of jail. The severity of a warrant dictates the strategy, and different times of year pose opportunities.

"If it's Thanksgiving or Christmas and it's a misdemeanor warrant, it's unlikely we're going to hunt that person down and take them away from their family," Lisko said. "However, if it's a violent felony warrant, you can be assured we're checking houses."

For the San Juan County Sheriff's Office, tactics are built on past successes.

"You have to think like a criminal to figure out where they're at and how to arrest them," Lisko said. "I was on patrol with a fellow deputy, and we pulled up the app to search our area. We found a felon who had four felony warrants for his arrest, some violent felony charges, and three known addresses. We started checking those addresses and found him at the third address, sitting in a convertible car with a tarp over it. The app helped us locate him, and although he did try to run and fight us, we were able to safely take him into custody without anyone getting hurt."

Keeping a positive reputation

Before they started using the app, deputies often made multiple visits to old addresses.

"Deputies had no effective way to log or share if a residence had changed to a different tenant," Lisko said. "They'd knock on a door, and the person answering would become so frustrated, saying, 'You keep coming here, and I keep telling you that I don't know this person.'"

The new app solves that issue, allowing a deputy to quickly flag and share a bad address with fellow deputies. This move supports the San Juan County Sheriff's Office in its mission to do good work compassionately. The team's philosophy is to partner with the community and uphold the law fairly.

"I've been out working traffic control at a crime scene, and people will yell 'thank you' out their car window as they drive past," Lisko said. "Those aren't the words a cop is used to hearing. It shows the rapport we've built with our community."

The next app project is designed to support crisis intervention—specifically for people with mental illness.

"We've gathered a database of people with mental health issues who submit information in anticipation of a crisis event. This data details the name of their doctor, the medicines they take, what gets them angry, and how to calm them down," Lisko said. "We're working to put that detail on a secure map so that deputies can see if the address they're responding to is the known residence of someone

In the summer months, Connie Mack Baseball games help increase the number of people coming into San Juan County on the weekends.

with mental illness. The data can mean the difference of a person calmly getting into our squad car to make a trip to the hospital versus having to fight them and taser them and take them to jail, or worse."

Lisko said the officers hope to repeat the success they had in deploying and using the warrant app.

"This is one of the few pieces of software I've released in my career that has been openly embraced with minimal to no complaints," Lisko said. "And it helps keep wanted people off the streets."

Fit to be featured on television

When producers for A&E Networks went looking for police departments to feature on its television program, *Live PD: Wanted*, they were drawn to the reputation of the San Juan County Sheriff's Office. They liked how the police department interfaces with the public and actively pursues criminals.

The day the new warrant app rolled out for San Juan County, A&E was there to start production of its *Live PD* show.

"The release just happened to coincide with the *Live PD* filming," said Lisko. "It was a happy coincidence the app made it on the show."

Since then, the app has caught on with sheriff's deputies.

"The next step is to roll this out to the other police agencies in San Juan County and with dispatch," Lisko said. "All the police agencies in San Juan County play nice together, and it makes sense for all of us to have access to the app to see all the warrants."

A version of this story by John Beck titled "San Juan County Sheriff Deploys App to Track Fugitives" originally appeared on the *Esri Blog* on February 26, 2020.

ADOPTING DIGITAL APPS TO SECURE THE STATE FAIR

Illinois State Police

NO ONE IS IMMUNE TO HOSTILE AND CONFRONTATIONAL comments from internet trolls—not even the police. Like many law enforcement agencies, the Illinois State Police deals with trolls who try to provoke individual troopers and the entire organization.

When one troll, well-known for threats against police, posted on social media sites about plans to attend the Illinois State Fair, a text alert went out to all troopers working the event.

"A couple days into the fair, a trooper saw somebody acting suspiciously outside the fairgrounds," said Nicholas Gray, disaster intelligence officer at the Illinois fusion center, named the Statewide Terrorism and Intelligence Center (STIC). "We quickly identified this man based on the notification and photo that was entered into the dashboard earlier in the week."

The new app-based approach helped the state police reduce known risks, as in the case of the threatening internet troll. The police increasingly use apps, in addition to their traditional radios, as a key means of communication for event security. During large public events, officers spend a lot of time identifying and restricting entry of people caught fighting, brandishing weapons, passing counterfeit money, drinking alcohol underage, and committing other illegal or disruptive offenses.

"We don't normally put intel out on the radio," said State Fair Safety Commander Lt. J. W. Price of the state police. "Instead, we would have a photo hanging on the wall at headquarters. With the app, we could push intel out to troopers."

Apps with photo and text alerts ensure that all troopers have access to shared intelligence. Adding a real-time map, with incident

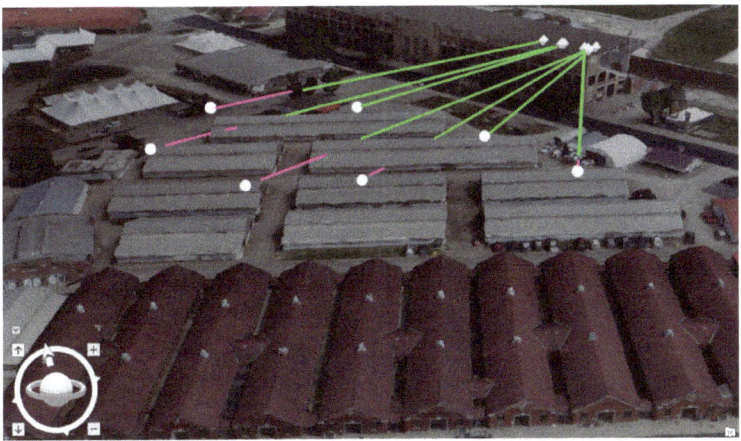

A 3D model of the state fairgrounds enabled the line of sight analysis that has become standard security procedure, showing areas where views are unobstructed from the tallest point, in this case the top of the grandstand.

and trooper locations, helps them coordinate efforts much more effectively.

Training opportunity

The state police have a long history of managing security at the state fair. The State Police Academy was located on the fairgrounds from 1922 to 1968, and rookie officers still often get their first experience working in uniform at the event.

The 11-day fair provides a good training ground for officers and new operational approaches. It attracts more than 30,000 people each day, with a lot of VIPs, including national political candidates. The fair also has a lot of areas that require security, including livestock competitions, horse and auto races, parades, a carnival, events such as tractor pulls and the demolition derby, and concerts.

"It has all the complexities of a multiday event," said Price. "It's also outdoors, so you've got the weather to deal with."

Because the state police handled security for the fair for so long,

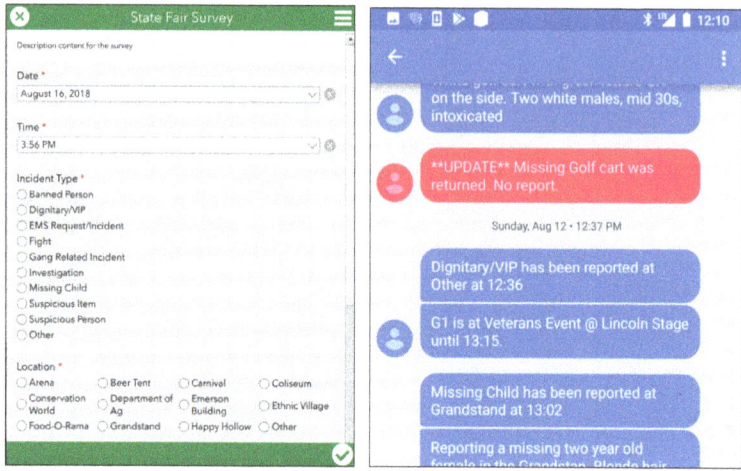

Troopers recorded incidents on their phones via a simple survey (left), quickly and securely sharing details with everyone on duty. Sample text alerts (right) provided condensed details that troopers could dig deeper into by viewing on a map.

they typically dusted off the standard binder, updated the layout and schedule, and proceeded as always. But after watching other agencies use real-time systems to manage event security, the state police decided on an apps-based strategy.

"We thought we could bring some efficiencies and some clarity to what people are doing," said Aaron Kustermann, chief intelligence officer for the state police. "We also thought it would be a good place to experiment where we control the facility and access to information, so we can try new ideas and new tools."

Making the app transition

Leaders at the STIC learned what peers had done for similar events using a GIS as the backbone for real-time situational awareness. The STIC already used GIS on a daily basis to help prevent, prepare for, and respond to disasters (both human-made and natural) and other public safety issues statewide.

The STIC had a good map of the fairgrounds, but the carnival layout changes every year. As the carnival setup neared completion, it flew a drone to capture current imagery for the map. STIC also added patrol sectors.

Fifty smartphones were provided by Verizon, and an additional 20 phones belonging to troopers were updated to add a suite of apps. STIC added ArcGIS Workforce for tracking the location of each trooper, ArcGIS Survey123 for quick and intuitive reporting of incidents using drop-down menus, ArcGIS Explorer for viewing incident details, and ArcGIS GeoEvent™ Server to send notifications to each phone whenever an incident was reported. In addition, ArcGIS Dashboards provided a one-glance view of incidents on large screens at headquarters and in the operations center at the fairgrounds to quickly update anyone on the status of the event.

"We learned from others that the simpler we can make things, the more widely they will be used," Gray said. "Literally all the troopers had to do was open the apps when they got on their shift to see what others had reported or to report something."

Officers embraced the platform quickly, Kustermann said. "Within an hour of the deployment users were using the application without any help."

Putting technology to the test

While the state police previously deployed tracking and some mobile applications, this was the first time they had a map-based view of troopers collecting and sharing information in real time.

The state police teamed with Verizon Wireless to ensure adequate connectivity to underpin this real-time capability. Verizon dedicated bandwidth through its Responder Private Core program, ensuring connectivity even when everyone at the event used their phones at the same time.

Ticket operators at each of the entry gates, and the EMS teams from the local Springfield Fire Department, also had phones with the real-time system.

"When there was a call for service, such as a medical emergency or other incident, we could look at the screen and see the closest trooper or EMS team," Price said. "We got a much quicker response time versus just calling who was assigned to a patrol area."

Testing and updating the system was a priority during the event.

"We made it clear to everyone that if there were issues or suggestions to let us know right away," Gray said.

"We changed the interface as the week went on," Price said. "I

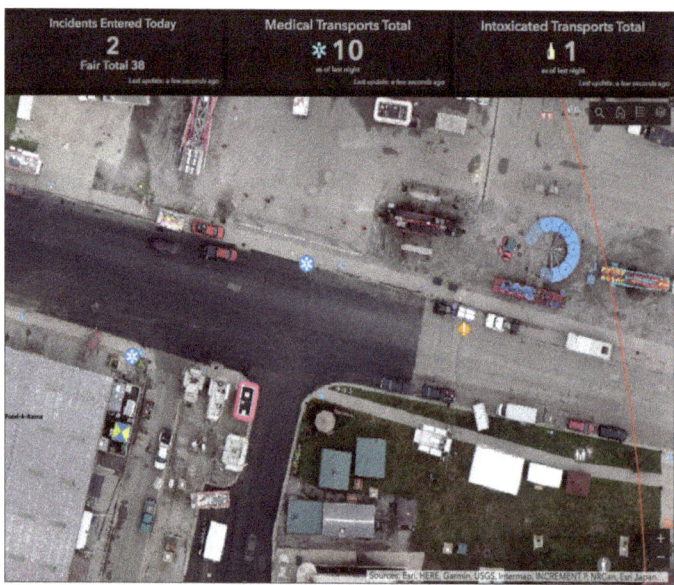

A dashboard view, including the drone-collected imagery of the fair setup, put incidents on the map for all to see.

would call to add this or delete that. There are still improvements that we could make, but I saw a lot of positive outcomes."

In the past, generating reports was a time-consuming manual task. With the apps, the flow of data automatically generated instant reports and an end-of-day incident tally.

Given the ease of trooper adoption, and the improved workflows, the state police plan to enact a real-time system for daily tasks and deeper investigations.

"Our agency has a leadership role during complex operations related to sensitive and unsolved murders, missing persons, and manhunts," Kustermann said. "We learned a lot about deploying applications, and it's definitely going to be part of the way we run criminal cases in the future."

This development was made possible through partnerships with the Illinois Terrorism Task Force, the Illinois Police Executive Administration at Western Illinois University, and the State of Illinois Department of Innovation and Technology.

A version of this story by Carl Walter titled "Illinois State Police Adopt Digital Apps to Secure the State Fair" originally appeared on the *Esri Blog* on November 1, 2018.

PLANNING AND MANAGING SECURITY FOR A TRAIN CELEBRATION

Ogden City Police Department

TECHNOLOGIES OLD AND NEW CAME TOGETHER IN downtown Ogden, Utah, culminating in a spectacular and safe Spike 150, the sesquicentennial celebration of the completion of the first transcontinental railroad.

Wowing the crowd of 5,000 people who gathered near Ogden Union Station on May 9, 2019, were Union Pacific Railroad's mighty steam locomotives Big Boy No. 4014 and Living Legend No. 844. These iconic engines brought train enthusiasts to Ogden from around the world. Many people followed the locomotives from stop to stop on their route to the city.

The two 1940s-era locomotives ceremoniously joined on the railroad tracks to mark the day—May 10, 1869—when the Central Pacific and Union Pacific Railroads connected at Promontory Summit in the then-Utah Territory. That connection opened the West to waves of settlers shortly after the Civil War.

"Thanks to the leadership of Abraham Lincoln and others, we have this transcontinental railroad to help us come together as a nation," Utah governor Gary Herbert said at the reenactment ceremony at a site just south of Union Station.

Behind the scenes, however, technology newer than steam locomotives—GIS services and software from Esri—played starring roles in providing security for the thousands of people who attended the three-day Golden Spike Sesquicentennial Celebration and Festival in Ogden.

The Ogden City Police Department used ArcGIS Online, ArcGIS StoryMaps[SM], ArcGIS Dashboards, and ArcGIS Survey123 to help manage and oversee various aspects of security, emergency operations, traffic, and even an event-related drone flight. ArcGIS was mainly used to help plan security for Spike 150 and provide real-time situational awareness during the event.

Forward thinking

Ogden is experienced in hosting huge events, such as the 2002 Utah Winter Olympics, but Spike 150 held significance in the city that has been dubbed the "Crossroads of the West." "Ogden is a railroad town," said David Weloth, director of the Ogden City Police Department Area Tactical Analysis Center (ATAC), which uses GIS extensively in its work.

A five-block area of downtown Ogden became the backdrop for the Spike 150 celebration. Public safety officials and organizers spent more than a year preparing for the five-day event. Historic 25th Street and another major roadway, Wall Avenue, were closed. The event footprint was divided into six operational zones, with a minimum of four officers present in each zone at all times.

Managing the security for the event fell on Ogden police chief Randy Watt and his team at the Ogden City Police Department. Watt looked to ATAC to assist in the preparation and management of the event. ATAC serves as the central nervous system for policing in the community, including surrounding agencies and the hundreds of thousands of citizens it serves. ATAC is staffed by a team of analysts, police officers, and commanders who monitor the city through a system of cameras in public spaces, sensor data, and other technology.

ATAC relies on Esri software and services for part of its daily overwatch responsibilities. For this event, Weloth realized that web apps could provide comprehensive operational awareness and field capabilities daily and during special events.

Planning began six to eight months in advance of the event. The

City of Ogden's GIS supervisor Josh Jones worked with Josh Terry, a GIS professional on the ATAC team, to create the apps. "Our enterprise GIS team did not have to be involved much," Jones said.

Using ArcGIS Online, ATAC mapped the locations of road closures, vendors, shuttle routes, parking sites, public cameras, portable toilets, aid stations, command locations, fire hydrants, barricades, the triage area, and bomb bunkers. Bomb bunkers are 55-gallon barrels where a suspected incendiary device (if found) would be placed, Weloth said.

All this information appeared on the Spike 150 Public Safety Map, which was created using ArcGIS StoryMaps.

Mapping all these locations was essential to creating a highly coordinated, transparent plan. "It helped us prepare more efficiently and effectively because we could see everything," Weloth said. "We were better prepared for this event than for the 2002 Winter Olympics."

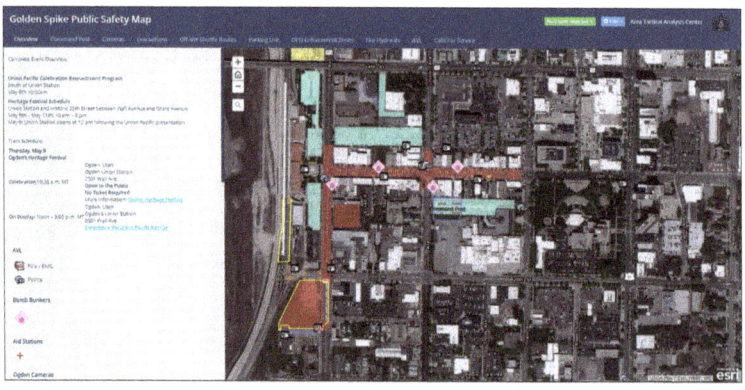

The Golden Spike Public Safety Map was created to share all plans for the day of the event.

Apps served many purposes

An ArcGIS Online map was embedded into an Esri Story Map Cascade[SM] app template to create the Emergency Operations GIS Viewer. Live dispatch calls related to the Spike 150 event were pulled into the map via ArcGIS Server, giving the team members in the Emergency Operations Center (EOC) a map with real-time updates on what was happening in the field.

Personnel working the event were assigned 12-hour shifts, and briefings were held twice daily, providing relevant updates. Priority data and voice connectivity made real-time situational awareness possible. Along with assigned public safety personnel, the Ogden Metro SWAT Team and the Davis County Explosive Ordnance Disposal team remained on-site throughout the three-day event.

Weloth and the team in the EOC used the mapping applications to help keep track of or shift resources when necessary, such as moving an officer from one operational zone to another. The maps also

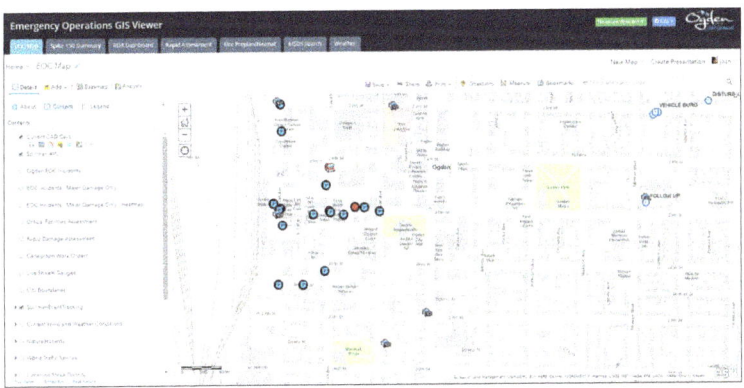

All data collected or reported to dispatch was displayed in the Ogden City Emergency Operations GIS Viewer. This is the tool used whenever the Emergency Operations Center is activated.

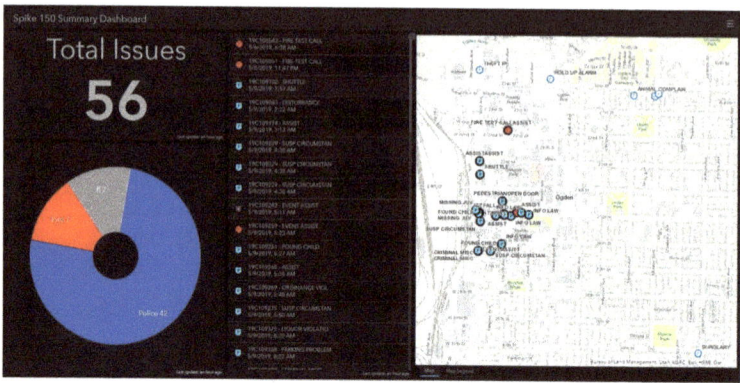

The Spike 150 Summary Dashboard.

displayed real-time locations of fire and emergency medical services (EMS) vehicles.

A dashboard created using ArcGIS Dashboards provided a snapshot of what types of calls officials were receiving in real time. As the event ended, the dashboard had recorded a total of 56 calls: 42 for the police department, 7 for the fire department, and 7 categorized as miscellaneous. Some of the calls included liquor violations, a foot pursuit, a lost child, and public disturbances.

Jones said the dashboard was used as an after-action reporting tool and also as a summary of real-time activity.

One of the apps created using Esri technology was public facing.

A traffic detour map was created using ArcGIS Online and embedded into the Visit Ogden website, according to Jones. The map gave residents and visitors the ability to see which streets were closed, the locations of off-site parking lots, and where they could catch a shuttle to get to the reenactment event.

Esri technology even played a role in an art project, according to Jones. An artist who was painting a watercolor of the meeting between Big Boy No. 4014 and Living Legend No. 844 in Ogden requested aerial imagery of the location where the two trains would come together.

"Requests for city drone flights are made using Survey123 forms and routed to the police department for approval via Microsoft Flow webhooks," Jones said. "Flight planning and record keeping are done via ArcGIS Online and [ArcGIS Dashboards]."

The dynamic nature of the Esri technology helped support the overall operation and ensure a safe event, Weloth said.

A version of this story by Mike King and Carla Wheeler titled "The Ogden City Police Department Uses Esri Technology to Plan and Manage Security for a Train Celebration" originally appeared in the November 2019 issue of *ArcWatch*.

SECURING THE WORLD SERIES WITH REAL-TIME TECHNOLOGY

Cobb County, Georgia

B Y THE TIME THE WORLD SERIES CAME TO COBB COUNTY, Georgia, in the fall of 2021, the technology and expertise demonstrated by the staff orchestrating public safety couldn't help but win the attention of Major League Baseball's technical team.

"They could walk into any conference room and with just a click they could see dashboards with maps showing the location of security staff, live traffic, and 911 calls," said Jennifer Lana, GIS manager at Cobb County Government. "They hadn't seen dynamic data that was so accessible before."

As the Atlanta Braves took control of the seven-game series against the Houston Astros, county officials wanted to plan for a massive victory celebration. But preparations had to be delayed because of a long-standing superstition in baseball that planning a celebration before the series ends invites bad luck. So when the Braves won Game 6 to win the series, Cobb County had 35 hours to organize a party for 350,000 guests.

The Braves had not won the series since 1995, when the team was in a different stadium and a different county jurisdiction, so there were no plans or knowledge about how to accommodate championship crowds.

However, Cobb County's investment in sophisticated mapping and monitoring technology and staff had been tried and tested during big events. That included home games of professional baseball's National League Division Series, the National League Championship Series, the World Series, and security for the National Football League's 2019 Super Bowl in Atlanta.

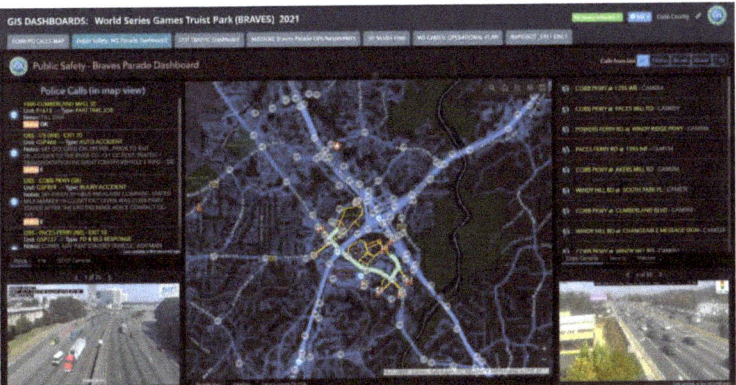

Public safety dashboard for the parade after the Atlanta Braves won the 2021 World Series.

This time, the huge celebration would include a parade procession that would wind through downtown Atlanta and culminate at the Braves' home ballpark, Truist Park.

Jeremy "J.D." Lorens, a lieutenant in charge of traffic management at the Cobb County Police Department, handled parade security using current maps and remote-controlled traffic lights. For home games, Lorens had already worked with a team of GIS specialists to apply mapping for crowd control, security, and traffic management. Now, Lorens needed the same detailed maps to create the parade route.

The GIS team provided tools to help the Braves organization and state and county officials define and approve the route. Lorens gathered barricades to match the narrow three-lane map he devised to reduce roadway width and put the crowd in proximity to players. "I knew there would be families with little kids," he explained. "I wanted to get them as close to the parade as possible."

During the parade, GIS maps displayed the location of every officer along the route, allowing incident commanders to see where each

Atlanta Braves players and fans celebrate during the 2021 World Series victory parade.

one was located in case of an emergency. The smooth operation of the parade allowed Lorens to craft an efficient plan for barricade removal when the procession was halfway through its route.

Advanced stadium technology

Before Truist Park opened in 2017, Cobb County authorities invested in advanced traffic management technology to move fans quickly and easily to and from the ballpark that sits near the intersection of two interstate highways. They wanted to make sure the stadium location improved rather than added to Atlanta's traffic congestion.

Lorens oversees traffic management at the ballpark and runs the control center with live feeds of 70 cameras monitoring pathways and "The Battery" entertainment district. There are 30 more cameras at critical intersections. GIS allows Lorens and the team to visualize the location and condition of every traffic light and camera. Using these inputs and controls, he and his officers synchronize the flow of vehicle and pedestrian traffic during regular games.

The live traffic dashboard helped public safety officials monitor increased traffic around the World Series games.

Cobb County's public safety departments use GIS dashboards and data on a regular basis to ensure citizen safety. These seasoned GIS users trust and know how to use the data, maps, and dashboards. They have also witnessed how location intelligence helps them. These trained users then became the trainers as more personnel were added during the playoff run.

For the 2021 World Series games, the number of police officers roughly doubled that of a typical game to handle larger crowds and longer hours. Officers were organized in three main groups: inside the stadium, traffic and operations, and overall campus security. The groups relied on GIS for real-time information and communication.

"We normally see 40,000 people in the stadium and another 5,000 people in The Battery," Lorens said. "For the World Series games we had 100,000 people on a 100-acre footprint, and we saw traffic start at 11 a.m. for a 7 p.m. game."

Location is also critical for the county's artificial intelligence (AI) system used to analyze foot traffic from the video feeds. GIS and AI technologies combine to inform changes in traffic-light sequences to keep people and vehicles flowing.

"We have 50 different datasets on the map from the footprint of the stadium, all the businesses, the lights and cameras, the traffic data feed, 911 calls, traffic from Uber and Lyft rideshare, and the location of Cobb County vehicles," Lana said. "We have built that awareness over time, so we didn't have to change much for the World Series, but there were modifications."

Lorens has learned that each game has a distinct traffic pattern, and he developed specific maps for day versus night games or weekday versus weekend games. When it came to the playoffs, though, he could not count on prior predictability.

"Once you get to a championship series, it doesn't matter what day the game falls on, because everyone takes off work," Lorens said. "You go to the most robust map you have, and you use that."

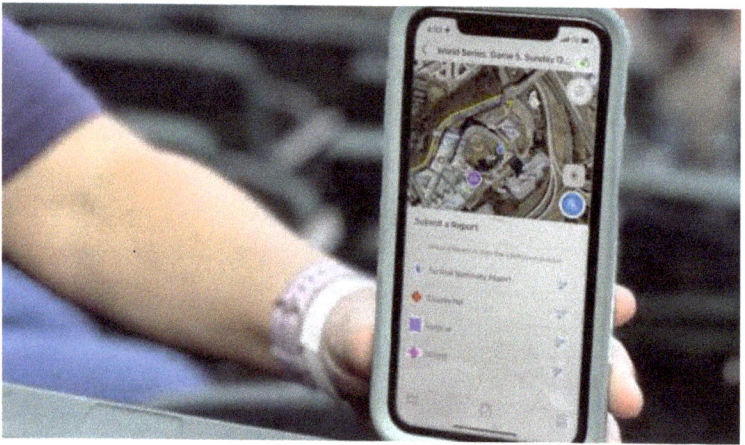

Undercover officers, who don't have radios, used a mobile app to see each other and where dangerous situations were happening.

Dashboards and real-time tracking

Having a well-provisioned stadium security setup helped Cobb County deal with the unanticipated nature of professional baseball planning. As the team worked its way through the playoffs, each subsequent win and series increased interest and attendance at the games.

The primary GIS toolset Cobb County relied on includes ArcGIS Mission; it tracked the location of roving plainclothes officers working throughout The Battery. With command-and-control software, the incident commanders in the operations center could make tactical resource assignments and get real-time updates from the field, and then share information across teams. That the app could be accessed on a smartphone was key to the stealth requirements of the operation.

"Undercover officers are in a unique situation because they don't have radios," Lana said. "But everyone on the planet has a smartphone, and nobody is going to think twice about someone looking at a map on their phone. On the app officers could see each other and where a dangerous situation was happening, and in the command center we knew no officer could get lost in the crowds."

Other smartphone features also helped.

"Our analysts," Lorens said, "could look up a license plate tag or provide background information. When it's just a text, that can get siloed between two people, and here everybody was able to see it and add to it if needed."

Putting the wraps on a stellar season

For Cobb County's World Series games, the use of Mission was tactical, guiding public safety response to incidents involving unruly behavior. "There were a lot of drunk people everywhere," Lana said. "Each time the Braves won, the crowds got bigger and crazier."

For Lorens and his public safety team, a map helped keep control of rowdy situations. "We were able to see where our people are and put them in places so everybody else saw them," he said. "For commanders, we could make sure we had coverage so that nothing critical happened."

Lorens and his team applied GIS in a more integrated way for the victory parade where the ability to track officers and floats in real time was critical, ensuring safety and easing traffic disruptions. It's a method the county will employ moving forward. "Now we've got a blueprint," Lana said. "I plan to build out the map of where we had bottlenecks and where we responded with more staff, and I'll keep that in my back pocket."

Fans in and around Atlanta hope there will be cause for another World Series and parade, another opportunity for Cobb County officials to demonstrate crowd expertise. But considering the age-old baseball custom against preplanning for victory, they won't be planning too far ahead.

A version of this story by Carl Walter originally appeared as "Cobb County Secures World Series with Real-time Location Technology" on the *Esri Blog* on February 1, 2022.

PART 3

CITIZEN ENGAGEMENT

GIS TECHNOLOGY SUPPORTS OPEN AND TRANSPARENT policing with tools to share information and engage the community. With a community engagement hub, agencies can share authoritative data including current and historic crime data, agency demographics and diversity data, and information about police use-of-force incidents. A hub can also promote community policing initiatives and help the public learn more about crime reduction efforts, report new problems, connect with neighborhood officers, and offer feedback about agency activities and strategies.

Publish

An open data portal provides law enforcement agencies with a quick and easy way to start sharing data with the public. Once an agency decides which datasets it wants to share, it can use GIS tools to publish key agency data such as crime and incident data, jurisdictional boundaries, agency demographics, use-of-force data, and other authoritative datasets. These kinds of public-facing applications make it easy for the public to get alerts about emerging crime problems, see how an agency's hiring practices reflect community demographics, and gain a deeper understanding of how and why police officers use force.

Connect

GIS can help enable community engagement and provide the community with tools to communicate outreach efforts, enlist the public's help, and brief members of the community on upcoming events or community policing programs that may affect them. Agencies can also improve access to neighborhood policing officers and provide tools for citizens to connect with an agency and share their comments and concerns about ongoing community problems.

Collaborate

GIS can also support collaborative initiatives and strategies to make communities safer. Many of the problems modern police face are more complex than simply responding to calls for service and can't be solved just by making arrests. Problems such as homelessness and opioid addiction are deep societal issues, requiring police to collaborate with multiple stakeholders to have a long-term impact. Police can use GIS as a collaboration tool among multiple stakeholders and can be used as a mechanism for understanding cross-jurisdictional problems by documenting incidents, sharing information, connecting people to resources and services, and measuring the effectiveness of response activities.

GIS in action

This section will look at real-life stories about how law enforcement organizations use GIS to promote transparency and accountability and connect with the community.

USING MAPS AND APPS TO STRENGTHEN COMMUNITY-BASED POLICING

Toronto Police Service

ON A GIVEN DAY, THE TORONTO POLICE SERVICE (TPS) receives 70 mental health crisis calls. A dedicated Mobile Crisis Intervention Team made up of a nurse and police officer juggle the demands of the crisis—they make sure the person doesn't threaten the safety of others or themselves and defuse the situation by connecting them to support services.

A new Community Asset Portal (CAP) application links the team to a list of services near their location. The app filters the data from Findhelp 211, a 24/7 helpline that catalogs services and providers, to make quick connections.

"The web application gives our officers access to relevant information in real time as they're interacting with a member of the

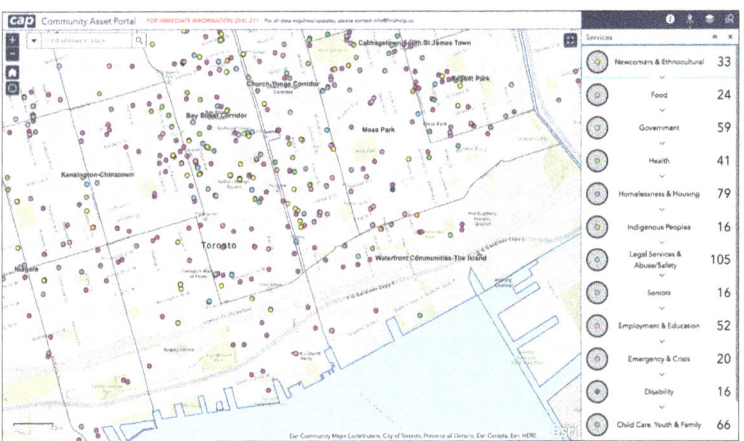

The Community Asset Portal displays data about community and healthcare services from Findhelp 211 for quick reference by officers and the public.

public," said Ian Williams, manager of Business Intelligence and Analytics at TPS. "Those accessing the portal at home can also know the resources around them."

Help in hand

TPS worked with a group of students from Ryerson University to create the CAP application. The students designed the app and connected it to Findhelp 211 data, gaining kudos for its ease of use with a simple location-centered interface. In addition to mental health facilities, the app locates meal services, housing, education, employment, transportation, and financial services.

"We're hoping that the referrals will lead to fewer calls to the police since the community can connect to resources to get proactive and not reactive assistance," Williams said.

"This unique partnership has increased collaboration between police and support services," said Sue Wilkinson, executive director of Findhelp 211 Central. "It has raised awareness of community supports and offered police 24/7 211 navigation support through the push of a button. Ultimately, this increases access to services for those in need, and makes the difficult job of community policing a bit easier."

The CAP application is one of many solutions in a broader modernization push at TPS that focuses on community-based policing. Community revolves around places, so the solutions often include tailored map-based apps to guide officers in the field.

"We have a quick way to stand up map-based tools for unique business processes," Williams said. "We've created tools for traffic safety, organized crime enforcement, neighborhood safety, and individual investigations."

Modernizing policing

The increased availability of data and more easy-to-use technology has fostered an apps-based approach to law enforcement that builds on the long tradition of intelligence-led policing. Connecting officers to applications in the field, backed by data and ongoing analysis in the office, leads to more informed approaches.

"Our focus is on standing up tools to present information they need to know about the location they're in at that time," Williams said. "We can present all of the incidents that happened recently, or if they're coming in after days off, they can know what happened in the past four or five days."

The data-driven app-based approach addresses multiple levels of operations:

- It informs investigators about trends and patterns in their area, including access to details about individuals who are known to commit particular offenses.

- It helps officers understand the best resolution for each incident type by presenting a catalog of different approaches.

- It guides the officers to account for cultural sensitivities by providing an understanding of neighborhood dynamics so that the message and approach make sense for each neighborhood.

- It assists administrators in the allocation of resources, with records of how many officers and investigators are required for each specific task.

"We provide data and visualizations to show what's occurring in a neighborhood," Williams said. "That facilitates a discussion about what we're going to do about it, and then we maintain that view to know whether it has been addressed and to what extent our approach was successful."

Conduits for the community

Toronto, one of the world's most multicultural cities, includes 140 unique neighborhoods with populations that have distinct needs. TPS assigns officers to work with neighborhoods for the long term, building partnerships with community leaders and citizens to strengthen safety beyond the resolution of individual incidents.

TPS conducts ongoing analysis of neighborhood demographics, languages spoken, cultural backgrounds, countries of origin, education levels, and other variables. Officials want to make certain they understand who lives there to appropriately guide officer engagement. They also conduct analysis by comparison, matching patterns of crime in neighborhoods with similar cultural makeup to see if there are trends that might impact like areas of the city.

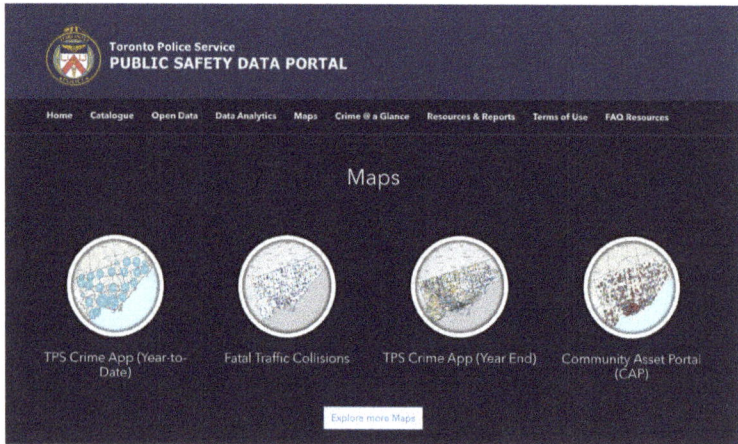

The TPS Public Safety Data Portal displays map-based applications including crime location data. Users can filter crime by neighborhood, major crime category, and date range.

The map-based Public Safety Data Portal provides an open data conduit to details on crime and the work that officers are doing in any one geography. The portal allows citizens to see their neighborhood or intersection, councilors to look at their wards, and members of parliament to look at their districts. The data is regularly refreshed, often in real time, to provide the authoritative view of public safety information in the city.

By providing the information transparently, members of the community and TPS share the same understanding of public safety challenges.

"The apps and mapping tools have helped us partner with other city departments and citizens within the city of Toronto," Williams said. "Because we're all responsible for the safety and well-being of the communities in Toronto."

A version of this story by Ryan Lanclos titled "Toronto Police Use Maps and Apps to Strengthen Community-Based Policing" originally appeared on the *Esri Blog* on April 9, 2018.

MAPPING CANADA'S MISSING CHILDREN TO QUICKLY REUNITE THEM WITH FAMILY

Missing Children Society of Canada

WHEN PARENTS ARE HIT WITH THE INDESCRIBABLE FEELING that comes with the realization that their child is missing, every moment counts. Fortunately, most of these events resolve quickly with the safe return of a child who may simply be hiding, was lost, or has run away from home. According to the nonprofit National Center for Missing and Exploited Children, which provides services in the United States, Canada, and the Netherlands, 94 percent of recovered children are found within 72 hours, with 47 percent found within three hours.

These facts, however, do little to diminish a parent's panic and dread that the worst has happened.

The best strategy to quickly return a child home safely is to get a photograph of that child in front of as many people as possible as soon as possible. A new application developed for the Missing Children Society of Canada (MCSC) puts those photos on a map—sending alerts to nearby citizens and agencies and gathering tips from anyone who may have information.

"When a child goes missing, they are at risk of being abused, sexually exploited, or becoming a victim of human trafficking," said Amanda Pick, CEO of MCSC. "The risk can be quite severe, and it can increase with each passing hour."

Confronting a common problem

In 2018, more than 42,000 children—115 per day on average—were reported missing in Canada.

"That's a shocking and sobering statistic," Pick said. "When

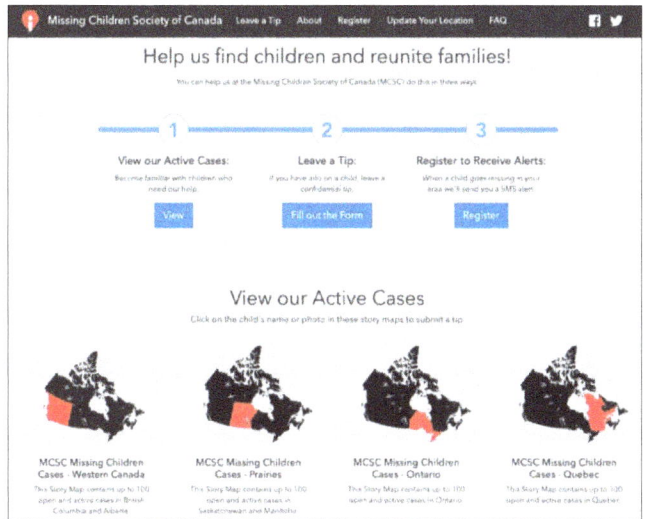

The MCSC rescu app provides three ways to assist with finding missing children.

people hear that number, there's a pause, because we simply don't expect the number to be that large."

Staff at MCSC teamed with developers from Esri Canada to create a web-based mapping app, called MCSC rescu, that uses ArcGIS to help reunite children with their families. Visitors to the app can see all active cases, with photos and details of children mapped to their last known location. The site includes a form that allows anyone to leave a confidential tip and add photos. It also provides a registry that people can join to receive an SMS alert directly from police if a child goes missing in their area.

"If a child goes missing under any circumstance, and the police immediately want to engage the community to help collect information, the tool provides the means to do that," Pick said. "We know that Canadians want to help, and the app gives them that opportunity."

Providing police with a powerful tool

The vision for this solution started with police and the desire to equip front-line responders with a way to engage the community. Police have the Amber Alert system, which issues messages to smartphones and roadway signs. But the threshold to create an Amber Alert is quite high; only 10 were issued last year.

"What about all the other children?" Pick asked. "Who are they, and how do we share that information? How can we help?"

Those questions drove the architecture for the app—a tool police can use to spread awareness across jurisdictional boundaries at the local, provincial, country, and even cross-border levels.

"Having pictures of a missing child, a potential offender, or a potential vehicle, that's really a game changer for us," said Cliff O'Brien, acting deputy chief of the Bureau of Community Support at the Calgary Police Service. "Every minute that goes by when there's a missing child is not only anguish for the parents, but it puts the child in more danger."

MCSC and Esri Canada designed the app to keep police in charge of determining which cases cross the threshold of a high-risk situation. High risk might include a child on medication or with a medical issue, a child lured away by a predator or at risk of abduction, or a child caught in a contentious custody battle, among other reasons.

The app gives police more control of the information released to the public. "We can release as little or as much as we need to. We may have other investigative strategies at play and may want information placed in specific geographies. We can push alerts to people in specific locations, even down to a certain street."

The promise of the app to create greater connections—using crowdsourcing to guard and protect the safety of children—has a lot of stakeholders excited.

"This technology will really help us communicate directly with any community anywhere in Canada where we think a missing child from our jurisdiction might be," O'Brien said. "The same goes for any small department all the way up to the Royal Canadian Mounted Police. It ties us all together from coast to coast."

Although the app's main job is to return missing children to safety, the possibility that it will also disrupt predation patterns is an added benefit, Pick said.

"It gets emotional when you start to think about the safety of the children you know," Pick said. "We all want to protect our own, our nieces and nephews, our neighbors' children, and all the children in our community."

Missing and murdered Indigenous women and children

In June 2019, Canada's prime minister Justin Trudeau spoke about murders and disappearances of Indigenous women and girls across Canada in recent decades, agreeing with findings that it was an act of genocide. An investigation had found Indigenous women and children are 12 times more likely to go missing or be murdered than any other demographic group in Canada.

"This history guides our actions," said Steve Burton, an inspector for the Tsuut'ina Nation Police Service. "The minute that we get a missing persons complaint, we immediately dedicate all our available resources to locating that person."

First Nations peoples often travel from reserve to reserve across broad areas and cross the border into the United States, particularly during the Pow Wow season of celebrations and dances in the summer months.

"With all those activities and events going on, it does pose a bit of a difficulty at times to identify and locate people who are on the move," Burton said.

Social media is often the first place investigators look at the start of any investigation.

"We look across social media—Facebook, Instagram, Snapchat, what have you—to see someone's last known communications and also to spread the word about the missing," Burton said. "Accessing all of these sites takes time, so having the MCSC rescu app as a dedicated platform that creates a quick alert, especially in our First Nations communities where it's an ongoing issue, could really help keep more children safe."

The app also fosters collaboration.

"One of the problems the policing community faces is the siloing of different agencies," Burton said. "In Calgary, the different detachments that surround our area are breaking down those barriers, communicating more, and working collaboratively with a regional perspective. The MCSC rescu app will further break down those barriers and will make our national partners more aware and involved."

A new ring road is being developed around Calgary that swings through the Tsuut'ina Nation, which will bring more people, and children, into the area. Here, police service has developed a close alliance across jurisdictions, aided in part by the MCSC rescu app.

"We are going to have a lot of citizens of Calgary and their children out here on the nation, and they deserve to be as safe as our children deserve to be," Burton said.

"Your race or where you live in Canada doesn't matter; we're looking out for all children," added O'Brien.

Enabling a digital transformation

The MCSC staff needed technology to bridge the gap between the few children whose circumstances qualified for an Amber Alert and the vast number of children reported missing. They engaged with Esri Canada to discuss and fine-tune the software architecture of the solution and in the process embarked on a digital transformation of the organization.

"We started by talking about communicating and sharing through geography," Pick said. "It soon became evident that the sharing was just the spot to start. The bigger opportunity was about making it into a system of support that has a real power to change outcomes."

Many MCSC investigators are retired police investigators.

"If you've worked a missing child or abducted child case that's never been resolved, I can tell you that those can haunt you," O'Brien said. "You'll always go over details in your mind, wondering if there's something that you missed."

In its focus on missing children, the MCSC staff tuned their system back end for internal reporting to organize staff workflows and deliver tools and dashboards for greater police awareness.

"When I understood that ArcGIS Hub^{SM} can power a smart city, it got me thinking that a city is only as smart as how it can protect its most vulnerable citizens," Pick said. "Combining the capabilities of data, collaboration, and support services that persist around a problem—there's just so much potential there."

Taking informed action

Law enforcement officers and MCSC staff have been urging people to sign up for the alert service in the MCSC rescu app to prepare for any local incidents.

"I see this playing out well in my jurisdiction in our 24-hour real-time operation center," O'Brien said. "If our incident commanders who are taking care of our city get a call, particularly if witnesses suspect an abduction, it allows them to flood the area with notifications within minutes."

The Calgary Police Service helicopter is equipped with a loudspeaker system for broadcasting alerts and descriptions over an area of interest. Adding the ability to share a photo and description with the community could help return a child to safety within minutes instead of hours or days as it is in some cases.

"The app includes a map that indicates how far someone could travel within a designated interval of time—whether walking or in a vehicle—from the place where a child went missing," O'Brien said. "It takes some of the guesswork out and communicates where we should set up our checkpoints or look for a vehicle."

While O'Brien cautions that the public should never go directly to an active crime scene, he notes they can take several steps when they receive an alert.

"We need people to look out their window, in their backyard, and around their property," O'Brien said. "We want them to have awareness and extra vigilance if they're driving to and from a store—keeping eyes and ears open. If we've pushed an alert to you, and you see the offender, phone 911. Be safe but try to keep your eyes on them, and we will respond to that area in force."

Spreading nationwide

The Calgary Police Service has pushed the MCSC rescu app to each of the 3,000 work phones officers carry. All employees are encouraged to sign up and ask their families to install the app on their personal phones.

The application has been endorsed by the Canadian Association of Chiefs of Police and various provincial-level police associations and city-level jurisdictions. The Royal Canadian Mounted Police have also taken an interest, with an agreement in place to work with MCSC to support missing children investigations using MCSC rescu and the network across Canada. Large corporations are also asked to spread the word and involve employees.

"We have put in a great deal of work already, but we know our journey is just beginning," Pick said. "There is an incredible opportunity to evolve and customize the app to provide really powerful change moving forward."

"I can't think of anything more noble than law enforcement and community coming together to protect our children," O'Brien said.

At MCSC, staff share an ongoing objective to empower children as well.

"We've been serving children and families at our organization for more than 30 years," Pick said. "Our work has evolved, but all our efforts activate in response to a child that has gone missing. There's an acute need to create a full circle where a child who needs us can let us know."

A version of this story by John Beck originally appeared on the *Esri Blog* on November 7, 2019.

APPS HELP PROVIDE TARGETED ASSISTANCE TO PEOPLE EXPERIENCING HOMELESSNESS

San Bernardino County Sheriff's Department

IN SAN BERNARDINO COUNTY, CALIFORNIA, THE LARGEST county in the United States, a growing number of people are experiencing homelessness. To address this issue, the San Bernardino County Sheriff's Department (SBSD) formed the Homeless Outreach Proactive Enforcement (HOPE) team to provide services to more than 1,800 people lacking housing within the county. With 20,000 square miles to cover, the HOPE team modernized its approach to homelessness by replacing spreadsheets with GIS to serve as the central system of record. Nine law enforcement agencies, an array of fire services, the US Forest Service, and several health and human service agencies contribute information to the GIS and use shared data to make informed decisions concerning homelessness.

Challenge

The county documented its unhoused population on paper during field surveys and manually entered the data into spreadsheets in the office. To find case information, county agencies cross-referenced multiple spreadsheets, a time-consuming and error-prone process. Because data wasn't shared between county services, such as the police department and health and human services, county agencies operated with incomplete information. The locations of unhoused populations were often anecdotal and not tied to an accurate map. This challenge made it difficult to connect underserved populations with the assistance they needed, such as mental health services, food, and shelter.

Solution

HOPE team members turned to ArcGIS Survey123 on smartphones and tablets to easily collect data in the field and quickly create detailed profiles of people without housing and contact records. This survey technique was expanded to other county agencies. They also used ArcGIS Collector to quickly create accurate maps of encampments of people experiencing homelessness. These maps enabled SBSD to understand the size and details of encampments, their distribution countywide, and their proximity to available services. As the team captured each survey and map, the information automatically entered the central system of record in near real time. Partnering agencies used ArcGIS Dashboards to track the progress and location of the field data collection efforts and monitor changes in the unhoused populations. Additionally, SBSD used these dashboards to understand how geography and the changing seasons affected people.

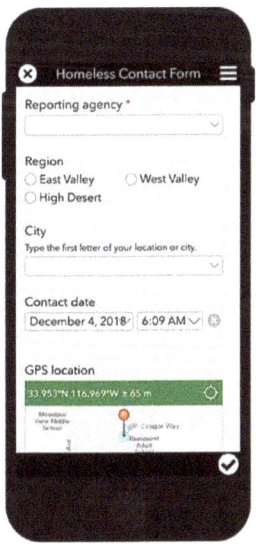

Survey123 app used by HOPE to quickly collect information in the field.

Results

The survey and map information collected by the HOPE team members helped them understand where to allocate the appropriate resources most effectively. GIS saved time, driving distance, and money. According to deputy sheriff Mike Jones, "With the [ArcGIS apps], we are able to provide a more targeted approach, saving us an additional hour to an hour and a half a day." The location-based system that SBSD created allows multiple agencies to contribute and search for homelessness records. It changed how the HOPE program serves these populations in the county and gave team members the ability to collect and share information quickly between agencies. Information sharing enabled the sheriff's department to collaborate with other police departments across the county, making it faster and easier to access contact information for people experiencing homelessness.

Having access to detailed digital maps on mobile devices gives officers the ability to collect data in the field and make it immediately available with real-time dashboards for faster decision-making and coordination of work in the field. Ultimately, GIS made it

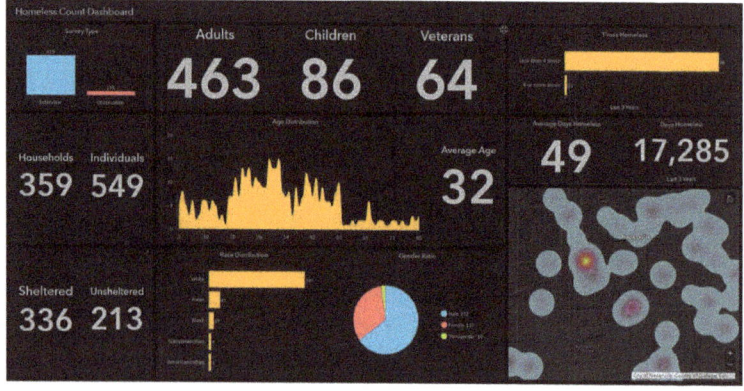

Dashboard used by the HOPE team to make decisions at a glance.

easier for officers, health workers, and people in other agencies to work together and connect underserved populations with the services they need.

A version of this story titled "ArcGIS Apps Help County Sheriff Provide Targeted Assistance to Homeless Populations" originally appeared on esri.com in 2019.

SPATIAL ANALYSIS OF OPIOID USE GETS LIFESAVING MEDICINE TO THE RIGHT PLACES

University of Tennessee at Chattanooga

ACCORDING TO THE CENTERS FOR DISEASE CONTROL AND Prevention (CDC), nearly 71,000 overdose deaths in the United States in 2019 involved opioids. In addition, the economic burden of the opioid crisis—including health-care and treatment costs, lost productivity, and criminal justice involvement—totals $78.5 billion per year.

Fortunately, overdose treatments are effective—particularly naloxone, a medication that quickly reverses the effects of opioids. Naloxone is easy to administer and has been used for many years as an immediate treatment for opioid overdoses. Distribution programs for naloxone, which train nonmedical persons to recognize opioid overdoses and administer the medication, have been around since the 1990s. And more recently, a widely adopted "leave behind" program expanded access to naloxone by having emergency medical services (EMS) personnel who respond to an opioid overdose leave behind a naloxone kit with friends or relatives in case the person overdoses again.

In Hamilton County, Tennessee, approximately 106 people died from overdoses and other drug-related issues in 2019, and that number increased by 50 percent in 2020 to about 160, according to the medical examiner (ME). Law enforcement agencies across the county responded to more than 700 overdose incidents in 2020, per the Tennessee Bureau of Investigation, and 90–95 percent of those calls likely involved opioids.

PART 3: CITIZEN ENGAGEMENT   83

When responding to an opioid overdose call, emergency personnel can leave behind naloxone, a medication that quickly reverses the effects of opioids.

The rise in opioid overdoses prompted area health organizations to search for a way to get the state's first leave behind program for naloxone up and running. Using GIS, a research team at the University of Tennessee at Chattanooga (UTC) and its Interdisciplinary Geospatial Technology Lab (IGTLab) analyzed opioid overdose trends in Hamilton County and determined the most effective way to distribute naloxone kits among local EMS stations.

Identifying clusters of overdoses

The IGTLab, run by UTC GIS director Charlie Mix, provides spatial data analysis and cartographic support to UTC faculty, staff, and students and the greater Chattanooga community. It also gives students unique learning opportunities that help them hone their GIS skills on real-world problems while supporting local nonprofits and other organizations.

"Over the years, the lab has provided GIS support in areas including conservation, outdoor recreation, and public health," said Mix.

In 2018, the IGTLab was approached by a coalition composed of the University of Tennessee Health Science Center, Erlanger Health System, and the UTC to help develop solutions to the area's opioid problem. The group had a limited supply of naloxone kits and wanted to develop spatially based criteria for allocating them.

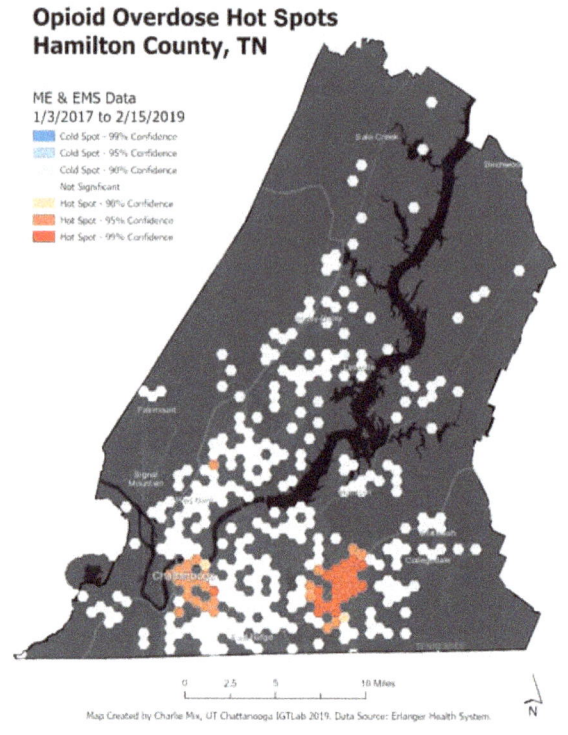

Identifying opioid overdose hot spots in Hamilton County, Tennessee, helped the research team determine where to target interventions.

Mix and UTC GIS analyst Nyssa Hunt used ArcGIS Pro, ArcGIS Online, and ArcGIS Insights to conduct the spatial analysis. They obtained two years of tabular EMS response and ME reports for opioid overdoses, which were scrubbed of personally identifiable information, from researchers in the coalition.

When it was time to geocode the data, team members processed the EMS data separately from the ME data, since they were in different schemas. Duplicate records in the datasets were identified and removed.

The EMS data and the ME data were then combined in ArcGIS Pro using the Merge tool, and data schema differences were reconciled using the Field Map parameter. This workflow resulted in three separate opioid overdose point datasets that were suitable for analysis: EMS data, ME data, and EMS and ME data combined.

The team then used the geoprocessing tool Optimized Hot Spot Analysis in ArcGIS Pro to identify spatial clusters of opioid overdoses for all three datasets. These represented areas in the county with the most overdoses over the study period, which was two years.

"This allowed [for] the easy identification of areas in the county with the most overdoses reflected by the data," said Mix.

Recognizing patterns of opioid-related deaths

To apportion the naloxone kits, the research team, along with the two doctors leading the study, decided to distribute them based on the number of opioid overdose calls each EMS station got. This would identify which EMS stations were likely to respond to the highest number of opioid overdoses and provide an opioid overdose map for EMS station coverage areas.

For this analysis, the team first needed to calculate the opioid overdose percentage for each EMS coverage area. This analysis would be used to determine the proportion of the total number of naloxone kits to allocate to each EMS station.

To evaluate the total number of overdoses per 911 response zone in Hamilton County, team members performed a spatial join of the point data representing opioid overdoses and a shapefile of the 911 response zones served by the EMS stations. Then, using the Enrich tool in ArcGIS Pro, they determined the total estimated population for each 911 response zone and used it to calculate the rate of overdose cases per EMS service area. This allowed the team to create maps that displayed the normalized rate of overdoses by 911 zone.

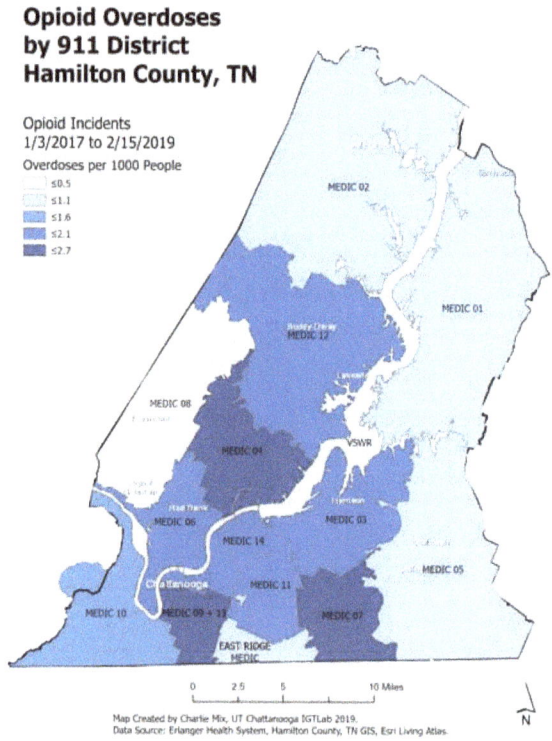

The team calculated the rate of opioid overdose cases per EMS service area and created maps, like this one, that showed the normalized rate of overdoses by 911 response zone.

"Visualizing the data in this way enabled us to recognize patterns of opioid-related deaths within the county," said Dr. Rebecca Martin, an emergency medical physician with Erlanger Health System who was involved in the project. "Identifying opioid overdose hot spots within Hamilton County allowed us to better understand the local epidemic and where to target interventions."

Additionally, because Erlanger Health System operates multiple regional hospitals and emergency departments throughout the county, the data has "influenced administrative decisions within the health system regarding where to initiate programs aimed at combatting the opioid epidemic in order to benefit the most patients," Martin added.

Gaining deeper understanding

In presenting their findings, team members from IGTLab also wanted stakeholders to quickly understand other descriptive data contained

Charts and graphics in ArcGIS Insights helped the team see the month, day, and time when opioid overdoses are most common and gain a better understanding of the victims' demographics.

in the EMS reports and ME records. So they used Insights to create and share graphs and other visuals that showed trends buried in the data, such as the age and racial characteristics of those who overdosed and the month, day, and time when these events occurred most frequently. It also gave researchers and medical planners the added benefit of interacting with the data and being able to perform their own analyses.

"The data has been used by administrators to reallocate EMS resources as well as emergency department resources to better serve the most vulnerable populations," said Jessica Whittle, director of research in the Department of Emergency Medicine at UTC. "In addition, some of our medics have indicated that it gave them a sense of encouragement because they felt that the high rates of overdoses they were seeing were everywhere, when in fact they were working in an overdose hot spot. Now they view themselves more as specialists rather than people fighting an unwinnable war."

Projects like this illustrate the power of maps and GIS as a collaborative tool for acting and bringing about change, Mix said.

A version of this story originally appeared in the Summer 2021 issue of ArcNews.

NEXT STEPS

The Geographic Approach to smarter policing

MODERN LAW ENFORCEMENT AGENCIES NEED LOCATION-based data to protect and engage the communities they serve. Many agencies are seeking new ways to improve services, increase their effectiveness, improve community satisfaction, and build public trust with their communities. GIS can be used to understand trends, increase transparency, and allocate departmental resources effectively. Here's how you can get started.

Identify foundational data

Gather and map your foundational data in your area. These layers include the basic infrastructure and administrative areas:

- Administrative and jurisdictional boundaries (city and county boundaries, police districts and precincts, public safety answering points, and so on)
- Crime, incident, and use-of-force data
- Population and demographics
- Public safety infrastructure (police stations, fire stations, and so on)
- Structures and structure type (single-family homes, multifamily units, commercial use, and so on)
- Major facilities and landmarks (schools, malls, places of worship, parks, stadiums)

- Health infrastructure (hospitals, clinics, assisted living facilities)
- Shelters
- Roads
- Bridges
- Dams
- Utilities
- Communications infrastructure
- Water features (lakes, streams, rivers)
- Parcels
- Addresses
- Street centerlines

Include hazard-specific data relevant to your area of interest. If you are unsure what hazards present the greatest risk, tools such as the Federal Emergency Management Agency's Resilience Analysis and Planning Tool in the United States can help you assess priorities.

Add ready-to-use, curated content from ArcGIS Living Atlas of the World, which contains several live feeds that provide dynamic, real-time information that can be used in addition to your local data:

- Weather feeds
- Disaster feeds
- Earth observation feeds
- Multispectral feeds

Discover more live feeds in ArcGIS Living Atlas at links.esri.com/atlas_live.

Also consider adding these and other real-time services for additional situational awareness:

- National Shelter System
- World Traffic Service
- WAZE Traffic

Identify data gaps

After collecting and organizing the base data, you can assess it and identify data gaps using a data drill. A data drill is a multiorganization exercise used to gain insight into how a community collectively thinks about, manages, shares, and uses data during a crisis.

Data drills are developed and conducted based on operational challenges involving data and are a valuable tool for disaster preparedness. Data drills can be designed around a specific scenario relevant to your community, such as an infectious disease outbreak, fire, or earthquake, to ensure you are planning for all data needs.

Here are a few suggestions for your data drill. Once the drill is completed, you can develop a plan to collect or create data where required based on these suggestions:

- Detail your organization-specific operational workflows and use cases based on the scenario.

- Identify the relevant decisions and determine what datasets, including metadata and data dictionaries, support those key decisions.

- Look next at your interagency workflows based on the scenario and identify key needs for data support.

- For each data point, identify the responsible organization contacts, roles, and responsibilities for this dataset.

- Identify whether you will need any data sharing agreements between partners, and start collecting and sharing the data identified in this drill.

Create and share maps

Once you've located the data sources you need, you can create a variety of maps to help keep your community informed and healthy, starting with these mapping tasks:

- **Map capacity:** Use the capabilities of GIS to map your facilities, infrastructure, locations of employees or citizens, medical resources, public safety resources, equipment, goods, and services to understand and improve access.

- **Map hazards:** Prepare for public health crises by creating maps showing the location of hazards that could potentially impact your community, such as areas with high crime rates, seismic vulnerability, flood zones, and wildfire risk.

- **Map vulnerable populations:** More specifically mapping social vulnerability, age, and other factors helps you monitor at-risk groups and regions you serve, which could be even more impacted in a public health crisis.

- **Share maps and plans:** Sharing your maps and plans with your community provides transparency, increases equity, helps citizens understand risks, and improves community preparedness.

- **Follow best practices:** Ensure that your maps and apps are ready to handle the load from the public and media during a crisis and that your GIS environment is ready for the next response.

Learn by doing

Hands-on learning will strengthen your understanding of GIS and how it can be used to improve law enforcement. Esri offers a collection of free online story-driven lessons that allow you to experience GIS when it is applied to real-life problems.

- **Introduction to ArcGIS Online:** Get started with web mapping with ArcGIS Online.

- **Combating crime with GIS:** Effective mapping enables law enforcement to detect and defeat crime at its source.

- **Assess graffiti incidents in your community:** Collect data on graffiti using ArcGIS Survey123.

- **Getting started with Police Transparency:** Learn how law enforcement agencies can use the Police Transparency solution to share authoritative information and communicate engagement efforts that build trust with the general public and community stakeholders.

- **Track crime patterns to aid law enforcement in ArcGIS Online:** Learn how to help a police department allocate resources to combat crime.

- **Identify crime patterns with data-driven policing:** Visualize and analyze crime incidents to identify areas with higher auto thefts to help recommend theft reduction strategies.

- **Getting started with Crime Analysis:** Learn how government agencies can use the Crime Analysis solution to conduct analysis.

Get there faster with GIS templates

ArcGIS Solutions for law enforcement reduces the time it takes to deploy location-based solutions in your organization. You can use these solutions, or templates, to take action that helps visualize calls for service, analyze crime patterns, increase transparency, manage operations, and more.

ArcGIS Solutions for law enforcement includes these templates and more:

- **Daily Activity Dashboard** can be used to visualize calls for service and incidents sourced from computer-aided dispatch or RMS and monitor daily activity.

- **Crime Analysis** can be used to enhance public safety, identify emerging trends, organize law enforcement operations, and plan crime-prevention strategies.

- **Police Transparency** can be used to share authoritative information and communicate engagement efforts that build trust with the public and community stakeholders.

- **Tactical Operation Planner (app)** can be used to create tactical operation plans for active shooter, barricaded gunmen, and other high-risk operations.

- **Tactical Operation Surveillance (app)** can be used to collect field intelligence that supports tactical operation plans at a given location.

- **Tactical Operation Dashboard (app)** can be used to access and review current and pending tactical operation plans.

- **Special Event Operations** can be used to develop safety plans and monitor public safety operations during a special event.

- **Traffic Crash Analysis** can be used to analyze crash data and identify streets and intersections where concentrations of serious and fatal crashes occur.

- **Public Safety Incident Maps (app)** can be used by law enforcement agencies to create a series of incident layers from computer-aided dispatch or records management data.

Learn more

For additional resources and links to live examples, visit the book web page:

go.esri.com/ptp-resources

CONTRIBUTORS

Matt Ball
Jim Baumann
Chris Chiappinelli
Mike King
Amen Ra Mashariki
Greg Mattis
Monica Pratt
Citabria Stevens
Christopher Thomas
Shannon Valdizon
Carl Walter
Carla Wheeler

ABOUT ESRI PRESS

AT ESRI PRESS, OUR MISSION IS TO INFORM, INSPIRE, AND teach professionals, students, educators, and the public about GIS by developing print and digital publications. Our goal is to increase the adoption of ArcGIS and to support the vision and brand of Esri. We strive to be the leader in publishing great GIS books, and we are dedicated to improving the work and lives of our global community of users, authors, and colleagues.

Acquisitions

Stacy Krieg
Claudia Naber
Alycia Tornetta
Craig Carpenter
Jenefer Shute

Editorial

Carolyn Schatz
Mark Henry
David Oberman

Production

Monica McGregor
Victoria Roberts

Marketing

Sasha Gallardo
Beth Bauler

Contributors

Christian Harder
Matt Artz
Keith Mann

Business

Catherine Ortiz
Jon Carter
Jason Childs

For information on Esri Press books and resources, visit our website at esri.com/en-us/esri-press.

www.ingramcontent.com/pod-product-compliance
Lightning Source LLC
Chambersburg PA
CBHW042321030526
44377CB00050B/761